Y0-BDB-927

HERO NEXT DOOR

DEDICATION

To HAZEL

who said I could do it and then proved her point by seeing that I did.

HERO NEXT DOOR

BY FRANK A. BURNHAM

AERO PUBLISHERS, INC.

329 Aviation Road Fallbrook, Cal. 92028

Library of Congress Cataloging in Publication Data

Burnham, Frank A
Hero next door.

1. United States. Civil Air Patrol. I. Title.
UA927.B8 358.4 74-81452
ISBN 0-8168-6450-0

Printed and Published in the United States by Aero Publishers, Inc.

FOREWORD

Born of crisis just six days before Pearl Harbor, the Civil Air Patrol performed a host of "Minuteman" missions during World War II. Then, after some shaky post-war months, when the organization came close to being consigned to the demobilization scrap-heap, the CAP won deserved support from the Congress and was made an official auxiliary of the Air Force.

"Hero Next Door" touches upon some of the incredible feats of the war years, but it is largely the peacetime, up-to-the-minute story of a uniquely American institution whose 60,000-plus members are dedicated to strengthening aviation and to making it safer and more understandable for fliers and the general public alike.

In tracking the story of the CAP down more than three exciting decades, the author has peppered his book with revealing anecdotes and has permitted many of the participants in his narrative to tell their own stories in their own words. And because he has been a part of CAP for so many years, he is able to relate the exploits of his huge "cast"—incidents that range from the heroic to the off-beat and the very funny—with a knowing touch and a keen eye for detail.

Having experienced more than a passing acquaintanceship with dozens of first-generation CAP leaders and rank-and-file members while criss-crossing the country researching the wartime story of the Patrol, I am gratified to be able to pick up the peacetime progress of this unsinkable, "grass roots" outfit through this book—and happy to know that the CAP story will be continued for years to come.

For a long time there has been a need for a factual, readable, up-to-date account of the accomplishments, frustrations and sacrifices of the men and women of the Civil Air Patrol. 'Hero Next Door" fills that requirement.

Robert E. Neprud

PREFACE

This is the story of the Civil Air Patrol told primarily in the words of CAP members themselves or those they have aided in time of crisis and disaster. It is a story of dedication and disaster, success and frustration, life and death. There was no intent in writing this book to provide a history, although it is historically accurate. Nor was there an intent to create a text book, although this work in its entirety is fact not fiction.

Simply, Hero Next Door is an effort to tell an absorbing story, one that needs to be told, in a readable and hopefully exciting manner.

In more ways than one, the "father of the Civil Air Patrol", Gill Robb Wilson, deserves credit for much of this story of CAP and its courageous men and women.

When he was editor of Flying Magazine, Gill Robb Wilson took an interest in my earliest attempts at magazine articles, offered professional advice and ended up publishing a number of my pieces over a period of years.

Later, as part of my Air Force responsibilities at Headquarters CAP-USAF I edited a department in Flying called CAP Wing Talk. During a visit to Gill Robb's New York office to coordinate a column, he suggested that since I was intimately familiar with CAP, I should write a book about it. His suggestion took root and later was nurtured by Col. Hal Basham, who was both my friend and superior during his tenure as Director of Information, Hq. CAP/USAF.

Essentially, much of the original research for Hero Next Door was done during the 1950s, at which time Hal Basham, Capt. Charlie Garnett, Lt. Col. Ross Miller and I ranged far and wide across the country covering major CAP stories on a first-hand basis as they were happening. Also, the exploits of CAP in World War II still were fresh in the minds of those involved.

Leaving CAP headquarters in 1958 to take an assignment as an Information officer with the then embryonic U. S. military space program, I maintained close contact with CAP and continued to assemble research material.

In 1970 while west coast editor of American Aviation Magazine, I was privileged to renew an earlier acquaintance with Bob Neprud, author of "Flying Minutemen," the first book about the Civil Air Patrol. Bob and I met frequently for several months and as we discussed CAP and how its wartime flying minutemen had matured into the versatile, public service organization now in being it became apparent a new CAP story was needed.

Actually, it took Col. Jack Ferman, until recently commander of the California Wing, to really get me "off my duff" and to work. Jack attracted me back to the program early in 1973 (I had been inactive for about 10 years) and put me to work as Director of Information for California. Simultaneously, work on this manuscript began.

The men and women of Civil Air Patrol in the 1970s are no less determined to make a significant contribution to their nation and their fellow Americans than those who rallied to Gill Robb Wilson's clarion call in the days of World War II. Day-in, day-out they prove their willingness to make whatever sacrifice may be necessary to get the job done. Some of the tasks they are called upon to perform entail a high degree of personal risk. Each year, a few more join the several hundred others who have made the supreme sacrifice. Those sacrifices, in fact CAP's whole contribution to society, take on even greater meaning when you consider the fact they all are civilian volunteers.

The man whose vision brought CAP into being died in 1966. At the time of his death, he was completing a book of his own chronicling his intimate experiences during the early days of aviation and his associations with its greats. Gill Robb's book was entitled "I Walked With Giants". It is significant that the final three words the father of CAP ever wrote for publication (in that book) were these—"Civil Air Patrol".

It is even more significant when they are read in the full context as he wrote them:

"As for my own part in conceiving and founding CAP, I can only note that I had a unique opportunity to foresee coming events and was in a responsible position to do something about it. If I did play some part, it was because I was surrounded by giants of great spirit who gave me their confidence and upheld my hands. To have walked proudly with them is all the credit I need or want. I had no thought of CAP as other than a Minute Man force to gain time for the military establishment. It was those who succeeded me who hewed out the greater destiny of the Civil Air Patrol."

If Gill Robb were here today, he would look upon the Civil Air Patrol of the 1970s, its senior members, cadets and general aviation members, with the same pride and pleasure that marked that earlier time.

And if Gill Robb were here today, he probably would take this manuscript, blue pencil in hand, and grumble at me:

"Well, what took you so long?"

Frank A. Burnham

Rancho Palos Verdes, California
June 1, 1974

ACKNOWLEDGMENTS

In addition to those individuals who provide an author with encouragement and moral support there are many who provide real and tangible assistance and without which no literary work would even survive its gestation period. I would like to extend appreciation to those who made a special contribution and in so doing offer sincere gratitude to all.

First and foremost comes to mind Lt. Col. Bill Capers, Director of Information, Hq., CAP-USAF, who coordinated the entire project from the Air Force and Civil Air Patrol point of view and who somehow got each chapter through review in record time; Ernie Gentle, the most helpful and understanding publisher an author could have; and my good friend, Harold Printup, whose creative work adorns the dust cover.

My thanks also must go to Maj. Gen. Lucas V. Beau, USAF Ret.; Maj. Shirley J. Bach, Chief, Magazine and Book Branch, Secretary of the Air Force Office of Information; William D. Madsen, Chief, Features Branch, Office of Information, U. S. Air Force Academy; and to Margaret Livesay and Frances Lewis, 1361st Photo Squadron (MAC)—the USAF photo depository.

Equally deserving of thanks are the many who answered my call for current information on CAP activities around the nation. Among those are CAP members Capt. Mary P. Tax, Washington Wing; Maj. Anabel Tucker, Minnesota Wing; Maj. Lindy Boyes, Hawaii Wing; Maj. Lou Dartanner, California Wing; Capt. Warren V. Huskie, Rocky Mountain Region; 2nd Lt. Dorothy Hinkle, North Central Region; 1st Lt. Carol M. Betterton, Virginia Wing; and Lt. Col. C. J. Seale, Puerto Rico Wing.

Last but far from least are my wife, Hazel, who critiqued, commented, rough-edited, proof read, indexed and glossaried; and my daughter, Ronley, who typed the original manuscript—all 320 pages of it—from my very rough copy.

TABLE OF CONTENTS

CHAPTER

Vintage 1940 Stinson swoops low over "survivors" during World War II training exercise. CAP crews were credited with locating and bringing help to hundreds of crewmen from tankers sent to the bottom by German U-boats prowling the Atlantic and Gulf Coasts during the early months of the war.

CHAPTER 1

That Others May Live

Alaska is a land of beauty and a land of magnificence. But like any frontier land, its beauty is steeped in savagery. Its magnificence is fraught with danger. There exists a constant battle of man against nature. For 24 agonizing hours, Charles Pearson, his wife Ola Faye, and their small children were the pawns in this battle for survival.

The Pearsons had been in Alaska for some time. This was to have been their fifth trip along the Alaskan highway back to the states by private plane. They had always traveled by private plane since living in Anchorage and had always taken the children with them. Little Joe, 4½, and Diane, 2½, already were old time air travelers. This, however, was the first flying trip for 7½-month-old Mark.

Charles, a dentist, was planning to attend the American Dental Association Convention in Miami and the Pearsons were to visit friends and relatives in Missouri, Georgia and California where Charles was to take some post-graduate work in dentistry before returning to Anchorage.

On a beautiful October afternoon, they climbed into the Cessna 180 and took off from Merrill Field. The weather was reported good ahead and they planned to land at one of the lovely lodges which can be found along the highway—either Gulkana or Northway, depending on how far they had proceeded by nightfall.

The CAA radio station at Gulkana still was reporting good flying weather. When the Cessna was perhaps 100 miles from Anchorage, the travelers had been admiring the beauty of Sheep Mountain and the glorious Alaskan sunset. Suddenly without warning clouds began to form. It began closing in ahead. When Charles attempted to turn back, he found it had closed in behind them. There was nothing to do but go on. Pearson called Gulkana radio and reported he was going on instruments. Gulkana specified an altitude and a heading and the Cessna began climbing to altitude. Charles told his wife to watch for anything she could see, but only for fractions of a second could she get a glimpse of the ground.

"I guess we had been on instruments about 25 minutes when it happened," Ola Faye related. "The clouds and fog apparently were less dense a few feet from the ground. Suddenly there before us we saw the trees and rocks of a mountainside. There was no time to think about it. Charles pulled back on the wheel to slow the plane up. Then we crashed.

"I don't think any of us were unconscious but we were stunned for a few minutes. The first thing I realized was that the children were crying but from the way they cried I didn't feel that they were seriously hurt. Joe had a small cut on his head and Diane complained that her leg hurt when she moved. At the time, I didn't know I had been injured at all until I tried to get up. Then my leg pained so that I couldn't move and I realized I would have to remain where I was until help came."

Charles' head had snapped forward in the crash and smashed against the instrument panel. His eyelids were swollen and his face was dark with bruises. He could hardly see but he kept telling his wife he didn't feel "much pain."

On the outside it didn't look as if any of the Pearsons were too seriously injured— all the seat belts had been tightly fastened. Despite the pain of his face Charles began trying to make the others comfortable. Fortunately they had all kinds of emergency gear aboard—in fact, all requirements for survival such as sleeping bags, flares, rations, flashlight and axe and also bottles of milk for little Mark. Ola Faye always felt there was too much emergency gear and admitted that occasionally she grumbled when they started on a plane trip because it got in the way. She felt differently now.

With dark the chill set in and the wind came up. Charles packed the sleeping bags around them. By this time his wife could only move her arms and about the only contribution she could make was to warm little Mark's bottles by keeping them close to her skin under the sleeping bag. Charles tried to find out where they were but by now it was too dark to tell.

"It was after 6 p.m. when I first looked at my watch," Ola Faye recalls. "We knew no one would find us this night but also we had faith in God and the fact that our fellow flyers would be out at dawn. Charles got in the back seat of the aircraft with me and we put Diane between us. We put Joe on the shelf behind the seat, wrapping him in baby blankets and afghans. I held little Mark. I was so grateful my arms were free."

"Charles had already tried the radio. It didn't work. There was nothing for us to do but wait. The doors of the plane had come partially off in the crash but even though we tied them in place the bitter Alaskan wind continued to tear them loose. It began to snow and it appeared that we might find ourselves in the midst of a full-fledged blizzard. But somehow we managed to keep warm."

Daybreak finally came. Charles took stock of the situation. It was still snowing and he knew that search planes would not be able to operate until the weather broke. Somewhere in the distance he seemed to hear the sounds of automobiles. His eyes were almost swollen shut by this time and despite the fact that he could hardly see he struck out toward the sound of the automobiles. At the time he judged them to be about five miles away. It was about noon when he left. It wasn't until much later however that Ola Faye learned what actually happened to her husband and how close he came to becoming a victim of Alaska's savage wilderness.

It was past noon when Ola Faye heard an airplane. Leaning over as far as she could, she looked out of the window of the Cessna. It was a

Navion and she learned later that it was a Civil Air Patrol plane piloted by Capt. Mason LaZelle with Senior Member Robert Husser as observer. Although she couldn't get out of the plane, she waved and waved as best she could from the window. There was no way of knowing if the searchers spotted the Cessna or not.

It was only a short time, however, before another plane came back and began circling the crash. It was W/O Garland Connell of the CAP and his observer, Senior Member William Christy. Connell dipped his wings and the injured mother knew she had been found.

It wasn't too long until several planes were in the sky over the downed Cessna including a helicopter. By mid-afternoon, an Air Rescue SA-16 Albatross amphibian reached the scene. The wind was blowing at better than 40 knots and yet as the injured woman watched, she saw a figure appear at the door of the rescue plane and plunge out. She lost sight of the parachutist and it wasn't until almost an hour later that he appeared at the window of the plane with the words "I am a paramedic—can I help you?"

"He asked if I was in pain but by this time it didn't matter," Ola Faye recalls. "I could stand anything. The paramedic, TSgt. Frederick Springborn of the 71st Air Rescue Squadron at Elmendorf AFB, and his partner, TSgt. Robert T. Elliott, later told me that they had never jumped in a wind like that before. When they threw out their spotter chute to determine the wind drift, they never found it again. But seeing only one set of tracks leaving the aircraft and knowing that there were four people still in the crash they decided to jump anyway. They missed the ridge and it took them 40 minutes to climb to the crash scene through the waist-deep snow."

The two paramedics immediately tried to call the rescue amphibian back with their portable radios. Apparently they had either been damaged or were frozen so they stamped out signals in the snow calling for a large rescue helicopter. They gave the woman Codeine to ease the pain and examined her leg. The hip was broken. They then gave first aid to the children. They put a splint on Diane's broken leg and sealed up their smashed airplane as well as they could to keep out the cold wind while waiting for the helicopter.

When the chopper finally arrived it had to circle nine or ten times before landing because of the high winds. Dr. (Captain) Robert D. Dayton, Jr., USAF Medical Corps, and MSgt. Frank Sackrider, his paramedic, assisted the other two sergeants in moving the mother and children from the crash scene to the helicopter 150 yards up the mountain. It was a rough trip with the four men battling the deep snow as well as the underbrush. Sergeant Springborn later said the jump from the amphibian was the worst either had ever made, but the air was so turbulent they were getting airsick in the airplane and were glad to get out of it.

The rescue helicopter took the survivors to Palmer where Ola Faye's joy was complete. Charles was waiting for her. Although the rescue amphibian had spotted him at a small lake about two miles from the crash scene in mid-afternoon and had dropped him a message telling him to await a pickup by helicopter, he never received it. After resting, he struck out again toward the sound of automobiles in the distance. He didn't know then, but he never would have made it because between the crash and the highway, lay the Nelchina River.

This was where the Army joined the Air Force-Civil Air Patrol-civilian rescue effort. Three Army Hiller helicopters were enroute from Ladd and Fort Greeley to Anchorage and were requested by Gulkana radio to join in the search. All that afternoon, the three helicopters scoured the mountainside and the adjacent valley. Just as they were about to give up, a tiny glimmer of light was seen by 1st Lt. William D. Brandon of the Second Aviation Company.

He made a pass toward the spot and he was able to spot Charles in the dense woods where he cupped his last remaining match in the palms of his hands to give off a light.

The trees, the darkness, and the 40 mile gusts made landing impracticable, if not impossible, but Capt. William H. Cox, commander of the Helicopter Detachment, realized that Charles was suffering from shock and exposure and might not survive another night in the open. He ordered Lieutenant Brandon and Capt. Lee Rodawalt, flying the other helicopter, on to Anchorage. They were running low on gas. Although he, too, was running low he elected to try to reach Charles. A hairy landing and take-off ensued before Cox could proceed to Eureka Springs where he transferred Charles to a rescue helicopter for the trip to Palmer.

It wasn't luck that brought Mason LaZelle and his observer to the stranded Cessna. It was a high degree of expertise in the art of the air search and rescue developed over a period of many years by the dedicated group of volunteer, civilian pilots, observers, radio operators and ground personnel who make up the Civil Air Patrol, civilian auxiliary of the U.S. Air Force. Some CAP members have been at the air rescue game for 30 years or more, ever since the organization recorded its first saves in the early days.

It is hard to say just when the first CAP air rescue mission was mounted, but Capt. Norman Kramer, the CAP squadron commander in Alamosa, Colo., and his observer, First Lieutenant Art Mosher, must have been among the very first.

Midwinter 1942 and 24 hours had passed since the four-engine Liberator bomber had given up the ghost and crashed with a thundering roar on a wind-swept saddle near the crest of 13,000-foot Little Baldy in northern New Mexico. The blizzard slashing the area gave little comfort to the nine survivors. Those with lesser injuries did what they could for their comrades but the situation looked hopeless. Having cheated death, they faced the distinct possibility that they now would succumb to intense cold and biting wind.

Another night and day passed. One searcher, a Flying Fortress, spotted them but the food and medical supplies it dropped were snatched away by the wind and came to rest on the crags below the tiny saddle. The search mission commander faces a dilemma—ground parties couldn't hope to reach the scene in time. Helicopters of that period couldn't cope with the altitude and the swirling winds. It probably was in a fit of desperation that the Army Air Corps mission commander sought the help of the newly-formed Civil Air Patrol.

Kramer and Mosher shortly were winging their way to the crash scene. The engine of their tiny, two-place Taylorcraft sputtered and protested against the rarified atmosphere. The mountain currents racked the little plane. Still it plugged along, Kramer nursing it to the 13,000-foot altitude and swinging in over the saddle. There lay the shattered bomber. They circled in closer and allowing for wind drift dropped packages of food and medical supplies from an altitude of less than 100 feet above the wreck.

"Bull's-eye!" Kramer chortled as the packages ended up in the hands of the eager survivors. Waggling its wings in a "keep your chin up" message, the Taylorcraft headed back to the base for another load of supplies.

On their second trip, Kramer and Mosher took a closer look at the little saddle of land lying between two jagged peaks.

"I'm going to try and land, Art," Kramer told his observer. Without hesitation, Mosher replied:

"Okay by me."

Kramer dragged the area, his eyes measuring every mound and hummock of snow that might spell disaster and add two more casualties to the probable toll. Finally, fighting gusts all the way, he reduced the power and maneuvered the little plane toward a predetermined spot on the saddle. As the gear touched, Kramer cut the throttle. The Taylorcraft bucked to a halt. This heroic action did more to spur the lagging spirits of the nine stranded crewmen than the actual delivery of the badly needed supplies which was accomplished in six more trips that day. On each trip, the altimeter of the tiny lightplane registered 12,800 feet above sea level, probably the highest a lightplane had ever taken off in those days. The miniature airlift made it possible for the nine crewmen to survive until a ground team finally reached them the following day.

The selfless action of Kramer and Mosher became one of the legends of World War II and one of the many such heroic actions that have become the heritage of the Civil Air Patrol which now is well into its fourth decade of public service. Today the CAP flies more than 80 percent of all the hours recorded on air search and rescue missions in the United States directed by the Air Force Aerospace Rescue and Recovery Service (ARRS). It is considered the "right arm of rescue" and rightly so. Take that August day in 1954—

The telephone in the alert room of the 41st Air Rescue Squadron at Hamilton AFB jangled. For a moment it appeared as if the duty officer might come in for a one point landing on his nose as he jumped to answer.

He listened for a minute then barked a half dozen questions at the other end of the line jotting down the replies on a note pad. Turning to the sergeant in the room he said tensely:

"We've got a hot one."

In a matter of minutes, the crews of three SA-16 Grumman Albatross amphibians and a SC-47 were being briefed while a para-rescue crew loaded their equipment into the Gooney Bird. At dawn, they would be airborne for Salt Lake City to set up a search base along with an H-19 helicopter at Boise, Idaho to be diverted to Moab, Utah where a civilian lightplane with three persons aboard was reported missing.

At the same time, the big Grummans were getting ready to take off from Hamilton for the long flight to Salt Lake, the duty officer was making another phone call. In the time it takes for a long distance telephone call to bridge the Sierras and the Great Salt Flats, rescue's right arm was sent into action. The Civil Air Patrol also was on the way.

Upon notification that a mission was on, Lt. John Pickering, Operations Officer of the Red Rock CAP Squadron at Moab, called out his personnel and at the crack of dawn, a half dozen lightplanes piloted by men who knew the local terrain like the back of their hands were in the air.

A half hour after the first SA-16 departed Hamilton for the search area, CAP search pilot Ed Shoup and his observer sighted the wreckage of the private plane. The wreck was in a canyon near Pack Creak, 3½ miles from the airstrip where Pickering had set up an advance base for the mission.

Ground rescue teams and a doctor were dispatched immediately. At the same time, Pickering called Hamilton informing them that the situation was well in hand and that the ground team with medical aid would be on the scene within an hour.

While the Air Force was busy recalling their aircraft and slowing down the huge rescue machine thrown into action by the call for help, Pickering's men had reached the scene. The pilot and one passenger were still alive. The quick action of the Moab unit was credited with keeping them that way.

In a report on the mission, Capt. Merlin R. Owens of the 41st Air Rescue Squadron said:

"With the information available at the time the CAP unit and particularly Lt. Pickering, the Redrock Operations officer, did a fine job in prosecuting the mission.

"With Moab being located in the far eastern sector of the 41st ARS area of responsibility, it is gratifying that such prompt, efficient action was taken by this CAP organization."

Another CAP log entry was stamped "MISSION COMPLETED."

These missions aren't isolated ones. Just ask any rescueman and he'll tell you the CAP is always on the job in the United States and Alaska. Rescue's strong right arm was flexing its muscle in late

February 1957 when a lightplane piloted by a 65-year-old Camden, Ark., man turned up missing on a flight from Camden to Little Rock.

Before word could be flashed to ARRS at Ellington AFB, Texas, CAP planes together with aircraft of the Arkansas Air National Guard, State Police, the Arkansas Forest Service and the Arkansas Game and Fish Commission were in the air. Eleven planes, eight of them CAP, were airborne. CAP Maj. Robert Bell and W/O David Ronnie spotted the crash. Bell flew back to his base and led the ground party to the scene before calling it a day. Rescue crews were saved a long trip from Ellington.

A month earlier a Sussex County, Delaware, CAP pilot spotted the life raft of a Long Island-based jet near Ocean City, Md., paving the way for discovery of the pilot's body following a criss-crossing, two-state search.

More than a score of planes including CAP complements from Milford, Del., and Salisbury, Md., searched for the jet missing on a flight from Langley AFB, Va., to Suffolk County AFB, N.Y. Capt. Earl M. Herholdt sighted the T-33's one-man life raft five miles south of the Maryland resort city. Herholdt contacted his CAP search base in the air. Using an Air Force C-45 as an airborne relay station, the CAP ground operator called a cruising SA-16 of the 46th Air Rescue Squadron giving him the position. Minutes later, the Albatross crewmen were adding their eyes to those of Herholdt and his observer. The body was sighted shortly and a helicopter from Chincoteague Naval Air Station recovered it.

Maj. William C. Ooley was one of only two senior members assigned to the Ely, Nev., Cadet Squadron when, early in March 1960, the unit was alerted to search for a Piper Apache reported missing on a flight in the area. Initially, Ooley and Lt. John Barainca attempted to locate the missing aircraft by ground search. When their jeep patrol could not make it in the heavy snow, Ooley took to the air. On his second sortie, the major spotted the aircraft and survivors huddled nearby in sub-freezing temperatures.

In his debriefing after the find, Ooley said that he "reasoned the aircraft would attempt a landing at Ely since he was flying blind and Ely is the only major airport with an omni." Ooley figured the aircraft had to be in the higher peaks a very short distance from Ely.

By this time, the snow had stopped at Ely but the nearby mountain passes still were closed by low visibility. Ooley found he could not cross over the Shellcreek Range into Spring Valley where he planned to center his search. Instead, he had to detour some 20 miles south before finding a pass. Even then his search was hampered by low ceilings and high winds. Mountains in the area run 12,000 and higher. In most places the cloud ceiling was only 9,000 feet with 40-mile-per-hour winds.

"Many times," he reported, "I had to throttle down to 80 miles an hour because of the extreme turbulence. At times in the approaches to the closed passes, I estimated my drift as much as 50 miles an hour.

Soon the snow increased and the ceiling dropped and visibility became almost zero. It was impossible to continue further."

Ooley now crossed the valley toward Mount Myria and searched the western slopes up to the ceiling continuing along this range until he was some 100 miles south of Ely. By now the snow had moved another 40 miles to the south and his fuel was running low—another hour at the most. He crossed over into Steptoe Valley and decided to head back to Ely.

"After crossing the Kalamazoo Pass," he reported, "I noticed a long dark object in the snow. I was astonished to find it was a twin-engine plane. When I circled closer I saw it was an Apache and that a woman had come out of the plane. I was now down to about 80 feet.

"The woman waved frantically even after I circled twice and nearly shook my wings off to let he know I saw her. Then she lay down three times in the snow to let me know there were three casualties in need of medical care."

Realizing that the survivors would need emergency equipment at once and that his fuel was now down to only 45 minutes flying time, Ooley headed for Ely where he put a cadet to work alerting the authorities while he pulled together what he could in the way of supplies—sleeping bags, sandwiches from a nearby cafe, a large candle since a portable stove wasn't readily available. This, he reasoned, would at least help raise the temperature in the plane's cabin and prevent frostbite. Taking off again, he headed for the 11,000-foot crash scene and by the time he had again climbed to altitude the sun was setting.

"I circled over the wreckage to figure out my drop," he recalled, "I still had a terrific wind and I had to come in on a downwind run and make a vertical 270 about 75 feet above the wreck. It was the best I could do. I flew over the area while the woman laboriously climbed to the first drop which was food. I continued to circle pointing out the other articles which were grouped together."

All told, Ooley made five air drops to the victims all of them landing within 150 feet of the crash. The supplies he brought sustained the woman and three badly injured companions until an Air Force helicopter from Hill AFB, Utah, arrived with medical aid. Shortly before dawn, the next morning, a ground rescue party that included Lieutenant Barainca reached the scene and evacuated the injured, one with a broken back.

When an individual elects to don the blue uniform—the Air Force uniform with distinctive insignia—of the Civil Air Patrol, he does so knowing that sooner or later he will have the opportunity to be of real service to his fellow citizens. Sometimes, however, that time comes sooner and the citizens are sometimes very close to the individual, even members of his own family.

Such was the case of Senior Member Jack Bradley of the Colorado Wing's Pueblo Composite Squadron. Soon after becoming a member of CAP, Jack played a major role in locating and rescuing his own brother, Bill, and two others who survived the crash of their aircraft in the mountainous area of southern Colorado.

Pilot Bill Bradley was flying Pat Floyd, Alamosa city manager, and his sister, Betty, from Denver to their home when they ran into a blinding snowstorm. Attempting to fly through La Veta Pass in search of a break in the clouds, their yellow and gray Cessna Skyhawk struck the ground at the 9,000-foot level west of Russell.

When CAP Group 4 was alerted to begin search, Jack obviously was one of the first aloft. He also was the one to find the crash early in the morning of the third day. Landing his aircraft on a highway, he met a tracked vehicle on loan from the San Luis Valley Television Co., and continued on with the ground party to the crash scene where his brother and the others were found to be suffering from nothing more serious than frostbite and minor injuries.

Missing aircraft searches do not always end happily—with the location and rescue of survivors. In fact, the nature of airplane crashes is such that the incidents which leave survivors are in the minority. Still, it is important that the crash be located. Often the aircraft is carrying valuable private or government equipment or papers. Many times it is urgent for the next of kin that the legal status of the estate be settled as soon as possible. Often all the family's assets, those involved in a small business, for instance, can be tied up indefinitely. Perhaps the most pressing reason, however, is the alleviation of the emotional suffering of those waiting at home. These are reasons enough for CAP and other agencies to press a search over an extended period even when the chances of there being survivors is long past.

The six-day, four-state search for a chartered C-46 passenger plane carrying 37 Korean War veterans and a crew of six is a case in point. That search still ranks as one of the most extensive in the Rocky Mountain area. Ultimately, the plane was found by CAP Major Richard W. Burt of the Utah Wing and the find was primarily attributed to his outstanding detective work.

With an explosive impact the C-46 had struck an 8600-foot ridge of Bear River Mountain just west of Bear Lake in the extreme southeastern corner of Idaho—only a few miles from the Utah state line. Authorities speculated that the C-46 had run into trouble shortly after passing over Malad City Radio where it made a position report indicating it was flying at 13,000 feet—well above the peaks. The plane probably was turning back to its last check point when the crash occurred inasmuch as it was then traveling 180 degrees to its previous course, officials said.

Major Burt sighted the snow-covered wreckage in an area which already had been subjected to an intensive air search by scores of planes. The find climaxed almost a week of 24-hour-a-day activity on the search. When the aircraft was first declared missing, Major Burt took a ground interrogation team made up of Lt. Ernest C. Lubean and Lt. Richard O. Abbott, both Utah Wing CAP officers, into the area of greatest probability in search of leads.

During that first day, they interrogated farmers, service station attendants and local residents in the area of Laketown, Utah, and in several southern Idaho communities. Each bit of information was thoroughly investigated and relayed to the 41st Air Rescue Squadron headquarters set up at Salt Lake City airport to supervise the entire search. That night, they got a few hours sleep in the City Court chambers at Montpelier, Idaho, before they were routed out to spend the rest of the night investigating reports of flares.

The party spent the next day in the vicinity of Kemmerer, Wyo., and slept that night in the Kemmerer City Council chambers—until 2 a.m. when the state highway patrol relayed a report of another flare supposedly seen by the crew of a Union Pacific train near Cokeville, Wyo., about 50 miles away. Major Burt hurried to the location where he flagged a train going in the same direction as the one from which the previous report was received. Riding to the exact spot where the train crew had reported seeing a flare, the major satisfied himself that what actually had been seen was the light from a sheepherders camp.

The next day also was spent in Wyoming following up a myriad of leads which were unproductive. He returned to Salt Lake City that night and spent the next day briefing in detail the Air Rescue authorities before returning to Ogden where he reassumed his position as CAP mission commander for the search.

Flying his first air search mission of January 12, Burt followed a hunch and made a personal check of the Bear River Mountain area. His hunch turned out to be a good one. Rescue officers were amazed that the plane had been found at all due to the almost complete disintegration of the plane and the heavy snow which had obscured the crash.

More than 750 hours were flown by some 98 planes from four CAP wings on this mission. CAP had a total of 227 members active in the four-state mission with all activities coordinated through the Civil Air Patrol's own radio communications facilities.

The Utah Wing alone flew 134 sorties—231 flying hours—on the search. Much of the time inclement weather and high winds prohibited the use of light aircraft. A total of 40 aircraft were used manned by 120 CAP personnel.

On 83 sorties, Wyoming Wing pilots flew 166 hours. Twenty-seven men in 17 aircraft turned in the total time. One CAP plane was damaged in a forced landing due to engine failure. There were no injuries.

In Colorado, CAP put 19 aircraft into the air search flying 210 hours on 110 sorties. Forty-three CAP personnel were involved.

The Idaho Wing put 22 aircraft into the search flying 169 hours on 89 sorties. Thirty-seven CAP members took part in the mission.

These missions flown over a period of more than 30 years are but a few of those logged by the organization which during World War II earned the name "Flying Minutemen." Today the primary operational mission of the Civil Air Patrol is air search and rescue and

CAP members are strictly prohibited from wearing arms or taking part in military or law enforcement activities. It was vastly different during the early years of World War II when CAP's tiny lightplanes sortied out to sea from America's shores, bombs and depth charges slung in jury-rigged shackles beneath their wings.

CHAPTER 2

The War Years

In the years just prior to the outbreak of World War II, concern that the United States was sadly lacking in airpower began to mount among the more than 128,000 licensed pilots and nearly 15,000 aircraft mechanics.

There was substance for their concern. Despite the fact that aviation was conclusively proven to be a powerful weapon during the latter part of the first world war, it fared very badly in the hands of the military traditionalists during the post-war period.

In 1921, for instance, the serviceable planes on hand or in storage consisted of some 1,500 JN-4 Jennies for training, 1,100 Dehaviland DH4Bs for observation, 179 SE-5 pursuit planes and 12 Martin MB-2 bombers. The majority already were obsolete. By July 1924, the aircraft strength had fallen to only 1,364 planes of which only 754 were in commission.

General Billy Mitchell's gallant personal sacrifice, when in 1925 he sought court martial in order to dramatize the nation's precarious situation in military air capability, actually accomplished little beyond a paper reorganization of War Department thinking. In July 1926, the Air Corps Act was enacted by the Congress. It was intended to strengthen the concept of "military aviation as an offensive, striking arm rather than an auxiliary." It also authorized a so-called five-year expansion plan. But, in fact, the position of the air arm remained essentially the same within the War Department and the proponents of air power continued to see their proposals sidetracked due to "lack of funds." Even the emergence of the heavy bomber in 1935 and the subsequent creation of a GHQ Air Force—an organization separate from the support aviation assigned to Army units—did little to improve the situation.

Even though a major conflict was shaping up in Europe, the tug-of-war continued here at home between the proponents of air power and our military "old guard." The great hopes for a real air force excited by the men of the GHQAF and the advent of the then-experimental four-engine B-17 were frustrated for most of the period between 1935 and 1939. Of the types of planes the Army Air Corps had on hand as of September 1, 1939, only one, the B-17, flew as first line after Pearl Harbor and there were only 23 of those. The B-24 was hardly off the drawing boards. The B-18 was the standard bomber, the A-17 was the standard attack plane and the P-36 was standard fighter. These three standard models comprised 700 of the 800 first-line aircraft on hand and they all were obsolete. To maintain and support these planes, there were 26,000 officers, cadets and enlisted men in the Air Corps, only 2,000 of them pilots and 2,600 mechanics.

By comparison, the German Luftwaffe had from 50,000 to 75,000 aircrewmen and a total strength of more than a half million. Even Great Britain had more than 100,000 military airmen. In first-line aircraft the Germans had 4,100 and the British had 1,900 compared to our 800. Thus, it was in the period from 1939 to Pearl Harbor that the civil airmen of the United States found reason to be concerned. They knew that they comprised the only immediately available resource to bulwark the ranks of our almost nonexistent air force when the gauntlet finally was thrown down.

Many of these civil airmen joined the Royal Canadian Air Force and the Royal Air Force. Others joined one of the U.S. armed services. There were, however, hundreds who were not eligible for military service because of age, physical condition or some other reason, but who still had the desire to help and represented considerable aviation-related talent. These men and women were ready to endorse any plan whereby their aircraft and their training could be put to use. Unbeknownst to most of them, such a plan was already in the making.

As early as 1938, one of the nation's foremost aviation writers, Gill Robb Wilson—who later became editor of Flying Magazine and for many years was considered the "dean of U.S. aviation writers"—made a trip to Germany. What he saw as a reporter confirmed his worst fears. Even then German air might was formidable. On his return he took his findings to the governor of his home state, New Jersey. With gubernatorial approval, he then organized the New Jersey Civil Air Defense Services—an organization which would permit the state to augment its resources with the civil air fleet available within its boundaries. The New Jersey plan called for using light planes for liaison work and for patrolling uninhabited stretches of coastline as well as enhancing security measures for protecting vital installations such as dams, aqueducts and pipelines. In addition, CADS would aid in policing the airports and "fingerprinting everyone connected with light aviation." The program also had the approval of both the Civil Aeronautics Authority and the personal backing of General H. H. "Hap" Arnold, chief of the Air Corps.

In a matter of months, similar organizations sprang up in Colorado, Missouri, Alabama, Kentucky, Ohio and Texas and the Airplane Owners and Pilots Association organized on a national scale its "Civil Air Guard." It was, however, New Jersey's CADS that served as the blueprint for what was to emerge as the Civil Air Patrol.

In the meantime, the Federal government began to face up to the facts. A civilian flight instructor refresher program was initiated and ultimately the Civilian Pilot Training program was begun. Those trained in these programs were, however, earmarked for military service and still no solid plans were made to utilize the some 25,000 light aircraft then licensed in the nation or those pilots and mechanics who could not qualify for the Air Corps or the air arms of the Navy and Marines.

Early in 1941, a number of events took place which collectively accelerated the formation of the Civil Air Patrol. In April, a formal plan was presented to President Franklin D. Roosevelt for the mobilization of the nation's civil air strength. In May, the Office of Civilian Defense was formed with former New York Mayor Fiorello H. LaGuardia—himself a former World War I pilot—at its head. Immediately, the proponents of the CAP presented their plan to LaGuardia. He approved and took quick action to appoint a special civil aviation committee to polish off the rough edges. This included selecting aviation leaders in each state as "wing commanders" of the proposed organization and seeking the formal endorsement of the military. Hap Arnold had already paved the way for the latter. Now he appointed a review board of officers to determine the roles and missions of the Civil Air Patrol and recommended that Army Air Corps officers actively help in its organization and administration. As things got down to the wire, Gill Robb Wilson went to Washington to personally handle the remaining administrative details. October 1941 dissolved into November and then into December. LaGuardia, Wilson and others who gathered on December 1 for the signing of the official document that created the Civil Air Patrol as part of the Office of Civilian Defense could not know that even then a Japanese carrier task force was steaming toward the Hawaiian Islands. Neither could they foresee the incredible role these "flying minutemen" would be called upon to perform in the ensuing months.

The Japanese attack on Pearl Harbor brought the war home to many Americans, but to others the remoteness of the Hawaiian Islands lessened the impact. In the days immediately following the attack, residents of the U.S. west coast also became more intimately involved as rumors of approaching Japanese warships and aircraft were rampant. The expected attacks didn't materialize, however, and by and large the nation's western shores felt little direct impact of this new conflict. Not so the Gulf and east coasts.

For many months, German submarines had been carefully charting our shoreline from the Mexican border along the Gulf of Mexico turning the corner around the Florida keys and on up the eastern seaboard to Canada. This was the route of the huge, ocean-going tankers bringing oil to eastern refineries as well as those headed up the coast and the protected convoy routes across the northern Atlantic to Great Britain. Now those tankers were fair game for the U-boat wolf packs and almost immediately they struck with disastrous effect. For residents of Florida, Carolina, Virginia, and other beach cities, the days and nights became a continual nightmare. While they remained in relative safety, they watched horrified at the carnage.

Our Navy was spread so thinly along this 1,200-mile sea frontier that it could not effectively combat the submarine menace. Help from the Air Corps was scant, if at all. Its meager forces were needed elsewhere. All that was available to protect the lumbering tankers were a handful of antiquated subchasers, five old Eagle boats, three ocean-going yachts pressed into military service, less than a dozen

Coast Guard vessels, four blimps and an occasional airplane.

January 1942—a dozen tankers were sent to the bottom. Forty-two more were sunk in flames, their backs broken by Nazi torpedoes during the next 60 days. Soon, the War Department stopped releasing the losses to the press because the toll taken by the prowling submarines was considered a serious blow to public morale.

In Washington there were still those in positions of authority who scoffed at the idea that a rag tag bunch of civilian fly boys with nothing better than puddle jumper airplanes could make any contribution at all, much less deter Nazi U-boats. But, the voices of dissent grew still one by one as the horrible toll mounted. And there were those including Wilson and Air Corps Maj. Gen. John F. Curry, who had been named the first commanders of the Civil Air Patrol—that knew that these civil aviators, with proper support, could at the very least scout the coastal waters for subs and report them by radio. This way the few combat aircraft available could be more effectively used. Whether the advocates prevailed or those opposing the plan just saw it as a desperate last ditch alternative isn't clear today. But it is enough to say that the CAP got its chance. Initially, it was for a 90-day test period, but those puddle jumpers with their civilian pilots and observers made such an impact they remained on the firing line until the submarine threat was over.

At first only three coastal patrol bases were authorized—at Lantana, Fla., Rehoboth, Del., and Atlantic City, N.J. By August 1943, when the coastal patrol mission was concluded, there were 21 such bases populated by a wide variety of circa 1940 airplanes—Wacos, Fairchilds, Stinsons, Rearwins, Cubs, Taylorcrafts, Widgeons and even a Standard or two.

The Atlantic is not a friendly ocean even where it laps against the sunkissed shores of Florida. It is a grey, green, sometimes even black, angry body of water especially during the winter months. Even today with modern electronic navigation systems, emergency locator transmitters, life rafts, survival suits and much, much more reliable aircraft and engines, experienced pilots are loath to fly much beyond gliding distance of the beach. In the winter and spring of 1942, the fledgling CAP's handful of volunteers, male and female ranging in age from 19 to 81, faced this same ocean with little but plain guts to sustain them.

Innertubes served as life jackets and life rafts in those first days of coastal patrol. Navigation instruments were limited to the compass in most of the available aircraft. In a rare case, one of the larger planes had a radio direction finder. Their communications capability consisted in most cases of a single channel high frequency radio subject to every kind of interference known to radio science. A forced landing out of sight of land, another plane or a vessel held out little or no hope for survival. Still the puddle jumper air force winged out of its makeshift bases morning after morning in search of the wily U-boat.

Doc Rinker and Tom Manning were two of those early to volunteer.

Theirs was one of the first sightings of a German submarine and one of the most frustrating. It was these early frustrations that also turned CAP from a mission of reconnaissance to one of attack.

On a pleasant May day in 1942, Doc and Tom spotted one of the undersea killers off Cape Canaveral. About the same time the sub's commander, sighting the aircraft and not knowing it was unarmed, ordered a crash dive and sought to escape out to sea. In its haste to get away from the plane, the sub became stranded on a sand bar—a perfect target for the bombers the CAP crew called for by radio. For more than 30 minutes, Rinker and Manning continued to circle the sub and continued their calls for combat aircraft. Meanwhile, the U-boat skipper managed to dislodge his craft and disappear under the swells.

The frustration of this CAP coastal patrol crew was multiplied several times in the first weeks. The U-boats were out there and because they had practically gone unchallenged, they were prone to lie with decks awash, torpedo tubes just below the surface and ready for firing. In any number of cases, the little CAP patrol planes were able to follow the subs for considerable distance even after they crash dived at the sight of the aircraft. While this was of great help to the tankers—it threw the U-boats off stride and obviously saved lives and cargo—it was of little satisfaction to the aircrews. They wanted blood—Nazi U-boat blood.

It had been a major victory to get the CAP formed and recognized. Another tussle resulted when CAP and its supporters asked for and eventually received the coastal patrol role. But it was an even bigger battle to get authorization to make the Civil Air Patrol a combat element of the U.S. war effort. As in the earlier cases, pure circumstances dictated the decision. The War Department needed help with the submarine problem and the civilian air arm of the Office of Civilian Defense—the CAP—represented its only additional resource. Again they got the job and almost overnight even the smallest CAP patrol craft turned up with 100-pound demolition bombs slung under wings and fuselages on makeshift shackles. Larger planes like the Grumman Widgeon sported a pair of 325-pound depth charges. Ingenious mechanics fashioned bomb sights—that really worked—out of hairpins.

On the maiden patrol, out of the Atlantic City patrol base, Maj. Wynant Farr, a slightly chubby New York cardboard manufacturer, and his partner, Al Muthig were so equipped. Innertube life vests lay on the floor of the Widgeon as the pair made their first sortie out of sight of land. Only a quarter hour flying time from shore they spotted a shattered tanker which by some trick of fate, had not burned. The water surrounding the ship was dotted with swimming survivors and motionless bodies. Farr radioed for help and continued his patrol. No subs, but some thankful merchant seamen who would live to sail again.

It was Farr, however, who was destined to record CAP's first "kill."

Capt. Johnny Haggin was flying right seat in the Widgeon that day. Almost immediately after takeoff, a flash report was received that another CAP patrol plane had made "contact" with a submarine about 25 miles off the coast, but that the aircraft was low on fuel and could not press an attack. Farr, still only about 300 feet above the ocean, headed for the scene at full throttle.

As they approached the area reported by the second plane, the crew glued their eyes on the tossing grey-green Atlantic—but, no U-boat. Like it had occurred so many times in the immediate past, the submarine had escaped, or so it seemed. The Nazi commander did not, however, reckon on the persistence of the CAP team or their sharp eyes. Farr and Haggin had learned their lesson well in the cram briefings given by experienced military advisors. They recognized the shadowy shape of a submarine beneath the surface where its commander sought safety until this airplane went away like the first had done.

They radioed a report to shore that they had reestablished contact and then held a council of war in the Widgeon's cramped cockpit. One thing for sure, they hadn't the experience to eyeball the sub and estimate its depth. Their depth charges were set to go off at a submarine's periscope depth and they were positive the craft was cruising considerably deeper. Since the twin-engine amphibian had several hours fuel capacity, the thing to do seemed to be to play the waiting game, continue to stalk the U-boat until its commander came up to periscope depth to see if the coast was clear and then hit it with their pair of 325-pound "ash cans."

Eyes strained to the point of blindness, buttocks numb from being glued to their seats, shoulders aching from the contortion of constantly peering from the Widgeon's side windshields, Farr and Haggin shadowed the sub for more than four hours before they were rewarded by seeing it inch to the surface, the periscope cutting a fine wake across the sea. And none too soon, the Widgeon's fuel was nearing exhaustion.

With Haggin flying the aircraft and Farr playing bombardier, the airplane went into a shallow dive coming in behind the sub. Haggin pulled up and lined the plane up with the periscope wake at 100 feet. Using a lot of by-guess and by-golly Farr released the first depth charge. The Widgeon pitched and rolled as the unaccustomed weight under its right wing dropped away. A tight turn brought the sub back into view and they saw the charge plunge into the water just ahead of the U-boat.

With a geyser of water, the depth charge detonated. The sub's bow pitched up and out of the sea. Shock waves buffeted the Widgeon as it curved around for its second run. Haggin and Farr could see large quantities of oil bubbling up around the submarine's forward quarter. Whether it was the effect of the first depth charge or whether the U-boat skipper was just crash diving again to escape, the aircrew couldn't determine, but the sub was dropping below periscope depth.

The second depth charge pinpointed the oil slick and this time the explosion brought debris to the surface along with more oil. As the Widgeon circled the scene more debris floated to the surface. Farr and Haggin looked grimly at one another. They had just sent perhaps 50 men to a watery grave. Then, the grimness was replaced by elation. Here was one member of the Nazi wolf packs that would not claim another tanker or more allied lives. CAP had drawn its first blood. The puddle jumper pilots were proving their mettle.

Anti-submarine patrol for the Civil Air Patrol lasted from March 5, 1942 until August 31, 1943—almost 18 months— or until America's defense industry and its military flying schools began to cope with the needs here at home as well as overseas. Also CAP's very success at the difficult coastal patrol mission helped put them out of business, but not before 26 CAP pilots and observers lost their lives and seven sustained serious injuries. In all, 90 aircraft were lost during that 18 months.

On the other side of the record, CAP flew approximately 24,000,000 miles in coastal patrol, sunk two submarines and had several "probables" to its credit in 57 attacks. CAP crews spotted 173 U-boats and in a number of instances were credited with an "assist" for summoning military aircraft or surface craft to make the kill. On its 86,600-odd sorties over water, CAP crews brought help by radio for the crews of 91 ships and 353 survivors of submarine attack.

There were many tributes paid to Civil Air Patrol's coastal patrol activities during and after World War II. They came from men like Navy Admiral Ernest J. King, General of the Army George C. Marshall, Admiral Adolphus Andrews, and, of course, Army Air Force chief, General "Hap" Arnold. But perhaps the most significant tribute came indirectly from one of Hitler's high ranking naval officers. It was given begrudgingly under interrogation by U.S. Intelligence Men after the German surrender. The officer was asked why the U-boat wolf packs had abruptly been withdrawn from off America's coasts in 1943.

His reply was recorded: 'It was because of those damned little red and yellow planes!"

Coastal patrol certainly was the most glamorous of CAP's World War II missions and in the early days of the conflict rated the top priority. But, it was not the only significant contribution its civilian volunteers made to the war effort. In some areas, CAP aircraft were assigned to fly cover over critical facilities like pipelines, power lines, dams and reservoirs. Along the U.S.-Mexican border extending for some 1,000 miles from Brownsville, Tex., to Douglas, Ariz., the Southern Liaison Patrol was established. Although it was called a "liaison" patrol, it was in effect a border patrol and its mission was to seek out unusual situations which might mean enemy agents were attempting to enter or leave the country.

This, like coastal patrol, had its own kind of flying hazards. While the coastal patrol planes had to fly over water, those involved in the border reconnaissance had to fly low and slow—so low that they could read the license plates of suspicious vehicles and even describe their occupants. These patrols also kept an eye out for suspicious aircraft. Private flying was prohibited. Only the CAP, the military, the commercial airlines or aircraft chartered by the government were authorized to operate. By reporting the type of aircraft and its flight path to ground stations, CAP crews could provide leads to clandestine activities which at best might only be illegal, but at worst could be in support of espionage or sabotage. The Southern Liaison Patrol also turned up other surprises. Wheel tracks leading to a "deserted" building were part of one routine report. When the lead was checked out by authorities, an enemy radio station was found.

CAP racked up another 30,000 flying hours during World War II on this border patrol mission and was credited with reporting more than 7,000 "out-of-the-ordinary" activities. Quite a number of them turned up the real thing like the pilot and observer who reported such an accurate description of the occupants of a suspicious car—they flew so low they could determine the color of the shirts and ties—that authorities at the border netted two enemy agents trying to get out of the country.

Once the word got around that civilian pilots and their aircraft had been organized to assist the armed services at home and were available for missions their small planes were capable of performing, a wide variety of opportunities were available. One of these came from the commands responsible for training anti-aircraft gun crews for service along our coasts as well as for deployment to combat theaters. A request from Fort Sheridan, Ill., got CAP-s tow-target mission going and an Evanston undertaker, Lt. Frank Hlavacek became the first CAP pilot to take on the task of dodging machine gun bullets and shrapnel. His squadron commander, Maj. Jack Vilas—one of the nation's pioneer pilots—was in the back of the red Waco handling the 1,200 feet of line which connected the airplane with the makeshift cloth banner trailing behind. Dye-colored ammo recorded hits on the target as the gunners, only 2,000 feet below, blasted away, often much too close to the little Waco, for three long and anything but dull hours.

It wasn't long before requests for tow-target missions spread to many other parts of the country where the coastal anti-aircraft batteries needed practice or where gunners were being trained. A number of different types of missions were ordered—flights in which the ground installations actually fired at the target like those at Fort Sheridan; missions where the gun crews merely tracked the target aircraft as it flew a predetermined course, and a predetermined altitude; and high altitude missions for the newly-developed radar-directed gun installations. No actual firing was conducted on the latter either, but it was necessary for CAP's small aircraft, usually without oxygen for the pilots, to fly over extended periods as high as 15,000 feet. Also included were night missions for search light tracking.

Even some of the low altitude tracking missions developed a few thrills. Such was the case when both the air crews and the ground gunners grew bored with the routine. Robert E. Neprud in his book "Flying Minutemen" which detailed CAP's wartime activities, tells of just such a mission.

During this, the pioneer phase of tracking for Army gunners, most of the missions were dry-runs over plotted courses. The towing of targets for actual ack-ack fire followed later. Until a regulation prevented CAP pilots from flying below 1,000 feet, the little planes would sometimes come screaming down into the very muzzles of the guns. Lieutenant John McGee of the Hicksville outfit did that only once, but it was enough.

McGee, a youthful, good-natured, devil-may-care Irishman, was on a dry-run mission over a battery of .50-caliber machine guns at Fort Totten, New York. He had a passenger with him this time, an Army captain whose job it was to coordinate the flight with ground operations.

After a couple of passes at 1,200 feet, the captain said, "Let's go down a little lower, Lieutenant."

McGee nodded, and the next pass was at 800.

"Hey, we're still too high," the captain remarked. "Drop down some more."

"Listen, Captain," answered McGee, "we're flying too damned low now."

"I still want to give these men some practice sighting on a low-flying plane. Let's go down."

McGee's mouth became grim. 'Okay, Cap, you asked for it. Hold your hat, here we go!"

The Taylorcraft plummeted out of the sky over New York Harbor and dived directly at the cluster of guns on the western shore of Long Island Sound. The gunners stared open-mouthed as the little craft raced down upon them.

This was too much for one of the men. He lost his head, and, instead of just sighting, he squeezed the trigger of his loaded gun, and McGee found himself staring into a .50-caliber in action. The bullets sprayed the little plane from stem to stern before he could pull up out of danger. They were lucky. A missing tail-wheel and a punctured fuselage were the total extent of the damage.

The lieutenant looked at his passenger. The captain, who had been so determined to dive on the guns, was a pasty white. He shook his head, "Never again."

CAP volunteers flew more than 20,500 target-towing and tracking missions. The work was dangerous—nearly as dangerous as coastal patrol—with seven CAP members losing their lives. Five more were seriously injured. The organization lost 23 aircraft.

Still another CAP World War II flying mission was forest fire patrol. Not only were the nation's natural resources critical to the war effort and it was considered imperative to detect and control natural fires expeditiously, enemy saboteurs also considered our forests a prime target. In announcing a plan for forest fire patrol to units through the nation, CAP National Headquarters observed:

"The danger of forest fires, whether by sabotage or by natural causes, must be guarded against with vigilance during this critical period. Timber is a strategic resource which must be protected. Fires in some forest areas would threaten power lines and other war facilities, and man-hours lost in putting out fires would be a drain on a much-needed labor supply."

Forest fire patrol became a major undertaking in many states with both the U.S. Forest Service and state forestry organizations entering into agreements with the CAP to provide this service. During 1943, for instance, in just one state, CAP flew 790 hours on 402 forest fire patrol missions and spotted 576 blazes.

Courier and cargo hauling emerged as one of CAP's most important wartime activities. In a two-year period, CAP's light planes moved more than 3.5 million pounds of priority cargo between military installations all across the nation. In addition, it provided an important means of moving specialized personnel from point to point where they were needed. The nation's commercial air transport industry found itself woefully behind the power curve when it came to meeting the challenge of burgeoning wartime industrial production and the growing needs of hundreds of scattered military facilities.

A 30-day experiment by pilots and planes of CAP's Pennsylvania Wing in 1942 proved conclusively that while the lightplanes often could haul no more than a few hundred pounds of cargo on a flight, CAP's planes were there when they were needed. For instance, critically-needed electronics parts to get a bomber into the air could be moved from a supply depot to the operational unit expeditiously and cheaply and without tying up another combat airplane.

CAP courier flights soon expanded until they were crisscrossing the U.S. with not only emergency supply deliveries, but with regular cargo and mail runs, between many remote military bases and airfields and the main commercial air transport hubs. In fact, airline operations experts took a page out of CAP's book when they created the feeder and air taxi lines in existence today. Where the big planes leave off, the little ones take over bringing air service to communities otherwise reachable only by surface means.

From the summer of 1942 until early in 1944, CAP's flying pony express carried 3,537,911 pounds of cargo and 543 passengers throughout the areas covered by the First, Second and Fourth Air Forces in the continental U.S.

The courier service also was not without its casualties. One of those was Lt. Margaret Bartholomew, then commander of the Cincinnati Courier Station. In October 1943, she and Lt. Melvin Myers flew from Cincinnati to Williamsport, Pa., to pick up and ferry another aircraft back. Lieutenant Bartholomew's story is one of those told so graphically by Bob Neprud in his chronicle of CAP's wartime service.

"Looks like a storm brewing," commented Myers as he and his

companion made their way to flight operations after landing.

Lieutenant Bartholomew nodded in agreement. "Yes, I noticed the clouds thickening toward the north on our way in. Let's take a look at the weather sequences. Hope they aren't too bad. I'm anxious to get home by tomorrow."

The sequences were of little comfort. A front was moving in over the city, bringing with it low ceiling, high winds, and continuous rain.

"Begins to look like you two will be sticking around for a few days," remarked the meteorologist on duty. "This stuff looks solid. It's going to hit in a couple of hours. All the flying that'll be done around here for awhile will be in the hangar."

That night the rains descended on Williamsport. The storm lasted four days, with Bartholomew and Myers twiddling their thumbs helplessly as they sweated out a break in the weather. The girl, anxious to be back on the job at the courier base, was particularly perturbed by the enforced delay. Finally, on the fifth day, the spirits of the stranded fliers soared. The rain stopped and the ceiling, which had remained at zero-zero for the better part of a week, rose beyond take-off limits. Bartholomew and Myers showed up at the airport bright and early for their trip back to Cincinnati.

The weatherman was still a trifle dubious. "I don't know," he muttered wearily. "I'm betting we haven't seen the end of this yet."

"Well, I've seen all I care to see of your fine Pennsylvania weather," the aviatrix laughed good-naturedly. "All I need is my clearance and I'll be back in Cincinnati this afternoon."

A moment later she waved the clearance at Myers and the meteorologist. "This is my ticket home. I'm on my way," she said.

"I'll be along in a few minutes, Margaret," Myers called after the girl as she went out the door. "The boys are getting my ship out of the hangar."

Myers walked over to the window and watched Lieutenant Bartholomew climb into the cabin of her Stinson. He saw the mechanic who had been warming up the engine jump aside and wave her away. Then the sturdy little monoplane swept down the runway and rose into the air, its nose pointed west toward Cincinnati.

The sound of the Stinson's engine was still in Myers' ears when the meteorologist called to him. "Here's the late weather coming in, Lieutenant. It isn't so good! Pittsburgh reports high winds and . . . " The weather had taken a sudden turn for the worse.

"Let's see that report." Myers' face whitened as he read over the sequence. Without another word, he raced up the steps to the tower and burst in on a startled operator.

"Get on the radio, bud! That girl in the Stinson who just left here is heading for plenty of trouble. We've got to get her back."

"Okay, Lieutenant." Then into the mike: "Stinson 865. Stinson 865. Stinson 865. This is Williamsport tower. This is Williamsport tower. Over."

Silence greeted the first call.

"Keep trying," Myers ordered, mentally kicking himself for allowing the girl to take off without waiting for that last sequence.

"Stinson 865. Stinson 865. This is Williamsport tower. Over."

No reply, only the crackling static.

The operator kept calling, again and again, in a constant but unsuccessful effort to reach the Stinson that was heading into the teeth of another storm raging in the mountains of western Pennsylvania. Finally he shook his head and put the mike down. It was no use. Either the plane's radio was out of commission, or the flier wasn't listening.

The rest of the story was told the next morning in the Williamsport SUN. A few hours after her take-off, Miss Bartholomew had encountered the full strength of the storm. A farmer in the western part of the state heard the sputtering engine of a plane above the whish of the wind and rain. He rushed out of his house and saw the storm-battered Stinson slipping and skidding downward in a frantic attempt at a forced landing. Suddenly the engine quit and the little craft went into a silent death-dive. There was a crash and then silence, as the Stinson plunged into the side of a small hill.

In addition to the 244,000 hours flown by Civil Air Patrol on anti-submarine patrol during World War II, pilots and crews recorded another half million hours in border patrol, tow-target and tracking operations, courier service, forest fire patrol, domestic security reconnaissance and search and rescue. In all, 64 CAP members gave their lives on active duty missions, six of those on air search mission. In this activity alone, the volunteer pilots and observers flew an estimated 50,000 hours from January 1942 to January 1946 and were credited with scores of "saves." In a single week during February 1945, CAP crews found seven missing Navy and Army pilots. This wartime activity served as the basis for CAP's primary operational mission today, search and rescue or SAR as it is called.

The red, three-bladed propeller on a white triangle superimposed on the blue shield as well as the red epaulets that adorn the familiar khaki uniform mark these Civil Air Patrol senior members as among the more than 100,000 who volunteered during World War II. Plane is an AAF Lockheed AT-18.

While their elders in single-engined lightplanes ventured out of sight of land seeking Nazi submarines off the Atlantic and Gulf Coasts, these CAP cadets got a taste of Army Air Force life at a Florida training base. Scores of young men who got their start in the cadet corps went on to carry the fight to the enemy's shores in AAF fighters and bombers.

A Civil Air Patrol show of strength during the waning days of World War II. Most of the contemporary (in that period) lightplanes shown here are highly prized classics today.

General Carl A. "Tooey" Spaatz, first Chief of Staff of the U.S. Air Force (center) poses with a pair of Civil Air Patrol's distinguished World War II leaders—Lt. Col. Wright "Ike" Vermilya of Lantana, Fla. (left), and Lt. Col. Harry Coffey of Portland, Ore. (right).

CHAPTER 3

REDCAP!

"ATTENTION ALL CALIFORNIA WING STATIONS, THIS IS WHITE BEAR FIVE SIX WITH REDCAP TRAFFIC."

Within seconds, Civil Air Patrol radio operators in all parts of one of the nation's largest states are at their transceivers. Audio gain control, UP. Pencil and paper ready.

It is the code word REDCAP that has galvanized housewives, teenagers, retired or self-employed businessmen and shut-ins to action. REDCAP means that a plane is missing and the Air Force has alerted the CAP to begin search operations and these are the people who insure that radio contact throughout the state is maintained 24 hours each day while most of California's 4000 members are at their jobs.

White Bear 56 is the base radio call of Maj. Lou Dartanner, 32-year-old owner/operator of a secretarial service/direct mail business in Santa Barbara. A CAP member for 12 years, she has specialized in SAR operations and admits to holding "practically every job necessary to a search base" before being made a Mission Coordinator. Subsequently, after some seven years as an MC, she was named one of the three California Wing Mission Control Officers who volunteer for 30-day tours of duty and during that tour, are on call 24 hours a day to the Western Rescue Coordination Center, McClellan AFB, Sacramento. This duty Lou shares with Capt. Bruce Gordon and Maj. Jerry Bollinger.

"THIS TRAFFIC IS REDCAP PRIORITY" Dartanner continues, "TIME ONE SEVEN ONE SIX ZERO ZERO ZULU NOVEMBER SEVEN THREE. ATTENTION ALL EMERGENCY SERVICES PERSONNEL. SEARCH MISSION FOUR ONE DASH THREE EIGHT THREE IS ACTIVATED THIS DATE FOR A PIPER CHEROKEE ARROW NOVEMBER ONE ONE ZERO ALPHA MISSING ON A FLIGHT FROM FALLBROOK TO CHICO. AIRCRAFT IS WHITE WITH RED TRIM. IT DEPARTED FALLBROOK AT TWENTY-TWO HUNDRED HOURS LOCAL ON A VFR FLIGHT PLAN ESTIMATING THREE HOURS PLUS TWO ZERO MINUTES ENROUTE. LAST RADIO CONTACT OVER BURBANK. SEARCH BASE BAKERSFIELD AIRPORT. MISSION COORDINATOR COLONEL PIERCE. BASE WILL BE OPEN TO RECEIVE AIRCRAFT IN ONE HOUR. RADIO CALL WHITE BEAR SIX. CHANNEL FIVE SIX, SEVEN AND SAR ONE TWO THREE DECIMAL ONE."

By telephone and through local area VHF radio circuits, wing relay stations are already passing the REDCAP traffic on to group and squadron commanders. At Shafter, LTC Addie Pierce (the wife of LTC Verne Pierce, commander of California Wing's Sector Charlie) is hustling to put together the material she must take to the airfield—the mission coordinator's kit, radio equipment, maps and charts, pencils, paper, first aid kit, etc. Some last minute telephone calls—the fixed base operator at the airport (standing arrangements permit quick reaction when a search is called); the Sector information officer to get a flash off to the wire services that a search is on (private citizens alerted to a search mission often provide valuable leads which point to the area of greatest probability); and to the local fire department. The latter is to her husband who is employed there and the message is succinct:

"There is a mission on so don't expect to find dinner on the table. Come to the airport and bring me a hamburger and coffee."

What happens in California, with only minor variations, is what happens in all 50 states, Puerto Rico and the District of Columbia when an aircraft is reported missing to the Air Force's Aerospace Rescue and Recovery Service (formerly the Air Rescue Service). The ARRS Rescue Coordination Center notifies the wing MCO when CAP help is needed. In some states, the rescue coordination center alerts the state director of aeronautics who in turn alerts CAP. In states small enough it is possible for the ARRS duty officer to directly alert a mission coordinator from a published list of those available. California, like Texas and Alaska, is big, however, and experience has shown that CAP mission control officers—equal in training and experience to the MCOs of ARRS—are a necessity and further take a load off the ARRS Coordination Center which has several western states to serve.

Any discussion of SAR requires a close look at the magnitude of air traffic in these United States. In 1971, for instance, the FAA reports that as of December 31, there were 133,870 active civil aircraft in the nation. Of these, 127,112 were fixed wing, 2,375 were rotor wing, 1,-607 were gliders, four were blimps and 74 were balloons. Of the total, only 2,698 were air carriers leaving 131,149 in the general aviation fleet. Searches for air transports are few and since rarely are search missions required for missing helicopters, gliders, blimps and balloons you must subtract another 5,000-odd aircraft. This leaves approximately 122,000 fixed wing general aviation aircraft operating from 12,070 airports across the U.S. In 1971, these aircraft flew more than 3.1 billion miles. To accurately consider the magnitude you must add hundreds of military aircraft flying tens of thousands of hours each year.

Since California is one of the most active SAR states for CAP, let's look at some more statistics. California is number 1 in the nation with 17,442 civil aircraft, 13.1 percent of all general aviation aircraft and more than 102,000 general aviation pilots—private, commercial, student, flight instructors, etc. From just seven representative general aviation airports in the state, there were 2,098,000 take-offs and landings in a recent year. Is it any wonder that in 1972, for instance, CAP's California Wing flew the equivalent of 14 times around the world on Air Force-ordered search and rescue missions?

Long before CAP becomes actively involved in a search mission, a number of other agencies have completed considerable spade work. Among them are local law enforcement units, the FAA and, of course, ARRS.

Since the National SAR Plan was adopted in 1956, the Air Force has held the responsibility for SAR in the so-called Inland Region—the 48 contiguous states, Alaska and the District of Columbia. The U.S. Coast Guard is the responsible agency for the Maritime Region off the coasts of the U.S., and that responsibility is assigned to Overseas Unified Commands in such areas as the Hawaiian Islands.

Within the Air Force, the overall responsibility for coordinating SAR is assigned to ARRS. ARRS maintains several operational squadrons in the inland Region, three of which are located at Eglin, Richards-Gebaur and McClellan Air Force Bases. Currently each of these units maintains a rescue coordination center, although there is a future plan to consolidate all three RCCs at Scott AFB, Illinois.

At the state level, the Air Force has a variety of resources upon which to call to augment the few specially-equipped, high performance aircraft operated by the ARRS squadrons. These resources include sheriff's air patrols, air units belonging to state aviation directors, military reserve units and private citizens. By and large, however, ARRS considers its primary local resource is the Civil Air Patrol.

In the most recent count, CAP had been accomplishing approximately 80 percent of all SAR flight hours in the Inland Region under ARRS direction. Even in those states where the office of the state aviation director is interposed between the CAP Wing Commander and the appropriate RCC in the chain of command, CAP ultimately performs most of the search.

Because SAR, in the final analysis, is so closely linked with the basic requirement for aviation safety, the FAA also takes a hand. Its responsibility actually precedes that of the Air Force since it is the FAA that is charged with implementing the national program of aviation safety. In this context, FAA operates the nationwide networks of aviation radio communications and navigational facilities and it is these facilities that first come into play when an aircraft is believed to be missing.

Basically, there are three ways in which the fact an aircraft may be missing is brought to the attention of the authorities—(1) it is on a flight plan filed with the FAA and does not close out the flight plan at its destination; (2) the pilot declares an emergency or the aircraft disappears from the radar screen while under radar surveillance and no further radar or radio contact is established; (3) a member of the family, a business associate, the owner of the aircraft, etc., reports the aircraft has not reached its destination. The latter is the case when the pilot disregards the recommendation of the FAA that a flight plan always be filed except for a training or pleasure flight remaining in the airport area and returning to its point of departure.

On a flight plan, the pilot reports in person, by phone or by radio to the appropriate FAA facility his intention to travel from point A to point B giving his estimated time enroute, his fuel capacity, his route of travel and details about his aircraft and himself which may assist in locating it if he encounters a problem which could cause an emergency.

When an aircraft is a half-hour overdue on a flight plan, the FAA checks with the departure point to see if he departed as intended, checks the Flight Service Stations along the route to see if the pilot indicated he was changing his flight plan or cancelling it, and takes action to forward the complete flight plan to the appropriate stations.

When the aircraft is one hour overdue, the FAA notifies the appropriate RCC and begins preliminary communications and ramp checks at airports along the route. At one and a half hours overdue or at the point of aircraft fuel exhaustion, FAA issues a bulletin to all stations (called an ALNOT), extends its communications checks and ramp checks to airports 50 miles on either side of the aircraft's flight route, and requests additional information from associates. If these comprehensive checks also are negative, FAA asks ARRS to begin search operations.

At this point, the RCC alerts the CAP and, through other resources including local law enforcement agencies, launches a minute ground check to determine if the aircraft might have landed at a field not having radio or telephone facilities and not checked during the initial ramp investigation. During this period, the CAP is selecting a search base and/or refueling bases, selecting a CAP Mission Coordinator who assumes overall responsibility for that mission and sends out the initial alert radio traffic to CAP units, throughout the state. Commercial communications are used to alert CAP forces not responding to the radio traffic.

A MAYDAY (the international distress signal) called by the pilot, disappearance from radar, an abrupt break in radio communications, or a report from the pilot that he is in trouble followed by no further communications causes the FAA to immediately alert ARRS. The standard methods of planning and beginning a search mission are implemented; however, in this situation local law enforcement agencies usually are the first to swing into action checking the immediate area where the aircraft was known to be when the in-flight emergency occurred or it disappeared from radar.

In the case of aircraft not on a flight plan, a relative or friend usually notices that a pilot is missing or overdue. This sometimes occurs a day or more after the flight should have been completed. Such notification sometimes comes directly to the FAA, ARRS or CAP, but more often than not is made to a local law enforcement official who relays the information to FAA and/or ARRS.

Extensive ramp and communications checks are made along the

route or routes the pilot may have followed. Friends, relatives, business associates and others who may have information as to the pilot's intentions are interviewed for leads and information as to the aircraft, fuel aboard, passengers aboard and the pilot's flying background. When there is sufficient information available to ARRS on which to base a search, CAP is notified and standard procedures go into effect. Sometimes, it may take 12 to 24 hours after the initial report before a full-scale search can begin.

While search and rescue as a role first came to the Civil Air Patrol, like its many other wartime missions, because there was no other organization with the resources to call on, it became evident during the immediate postwar years that this was a mission the CAP was peculiarly qualified to accomplish.

Military aviation was firmly established as a major element of the armed services and there were plenty of military aircraft available and crews to fly them. But they were large, fast and expensive. With the exception of relatively few small liaison planes used for artillery spotting, they did not lend themselves to the low-and-slow flying needed for air search. Also military pilots did their flying high, fast and for long distances for the most part and did not gain the intimate familiarity with every nook and cranny of the local terrain like civilian airmen.

Ultimately, when it was decided to make CAP an active auxiliary of the military, it was given direct mission to support the search and rescue requirement. World War II liaison planes like the Piper L-4 and Stinson L-5 first were loaned by the military to CAP. After the Korean conflict, these and additional aircraft of this type, the Aeronca L-16 and the Piper L-21, were given to CAP. In subsequent years, Cessna O-1 and de Havilland U-6 aircraft have been added to the fleet along with some light twins, the Beech C-45 and the Cessna U-3, and trainers such as the Beech T-34 and the Cessna T-41.

With the initial on-loan aircraft—approximately 500 in all—augmented by CAP-member-owned planes the organization flew 9,801 hours on actual searches requested by the USAF in 1952 upping this to 12,139 hours in 1953. Headlines like these began appearing across the country:

CAP SEARCHERS SAVE FOUR FROM ALASKA WILDS. PLANES FROM 5 NEW ENGLAND WINGS JOIN ALL-OUT SEARCH FOR BEECHCRAFT. WRECK SPOTTED AFTER ARS CLOSES SEARCH. EVERGLADES RESCUE—DOWNED PILOT SAVED. FIND WRECKED PACER IN TEXAS. CAP FINDS MISSING DOCTOR IN ALABAMA.

The stories with these headlines went like this:

"FAIRBANKS, Alaska — Four crash-shocked passengers of a downed light passenger plane were picked out of desolate Alaska wilds by alert Civil Air Patrol fliers who spotted the totally smashed craft and its luckily uninjured occupants less than 24 hours after the call for help was sounded. Pilot Hawley Evans and his observer, Mrs. Bob

Hunter, made the find in the midst of an intensive search which put an umbrella of military and CAP planes into the air over the forested regions surrounding Fairbanks.

"Quick action by the CAP searchers saved the crash victims from spending another night in the wilderness. Exposure to a second night of the Alaskan cold might have proven fatal to shock-weakened passengers of the ill-fated plane."

"PROVIDENCE, R.I. — Civil Air Patrol stations in five New England states were alerted May 19 when a two-engine Beechcraft was reported missing with three persons aboard. The intensive search which followed was resolved the next day when the body of one of the passengers was found floating in Block Island Sound, just off-shore of Watch Hill and four miles from the Westerly airport."

"MONCKS CORNER, S.C. — Remnants of a plane, identified as belonging to the yellow and black Stinson owned by Dr. H. S. Settle, missing since August 4, were found August 18, in Lake Moultrie. The search is continuing for the body of the doctor.

"CAP searchers persisted in the hunt even after personnel of the 48th Air Rescue Squadron of Maxwell AFB had concluded the official search activities after the customary eight days had passed without results."

"HOLLYWOOD, FLA. — A Hollywood man and his 12-year-old son, lost for two days and a night in the Florida Everglades, were rescued from the watery, insect-infested wilderness September 27 after being sighted by Civil Air Patrol fliers who had joined in the area-wide search for the missing pair."

"GENESEO, ILL. — Thirteen hours after he ejected himself from a crippled Navy Banshee jet fighter, Lt.(jg.) J. J. Meder of Pittsburgh, Pa., was spotted in a muddy soy bean field by Illinois Wing CAP searchers.

Lieutenant Meder was suffering from multiple fractures of both legs, internal injuries, shock and exposure. Physicians at Glenview Naval Air Station said he would not have lived another night. Although his condition is still listed as 'critical', he is expected to live."

"HUNTSVILLE, TEX. — The scorched wreckage of a private plane missing for three days in South Central Texas was spotted August 28 by Capt. Rube Thode, USAF-CAP Liaison Officer with the Texas Wing, flying over the Piney Woods area with members of the Tyler Squadron.

"The gray Piper Tri-Pacer was found on its back in a region sprinkled with pines and heavy vegetation. Killed in the crash was Pvt. Klien Evans, 21, Lake Charles, La., a soldier who had been enroute from his base in El Paso to Lake Charles."

"MUSCLE SHOALS, ALA. — An extensive air and ground search by CAP units from three states were conducted for a missing aircraft piloted by an Alabama physician.

"Dr. Gilbert R. Nelson was flying his Cessna 210 from Birmingham to Muscle Shoals on a routine flight back to his home. He had flown the route many times.

"Civil Air Patrol pilots from throughout Alabama took part in the search, flying nearly 200 sorties during the two-and-a-half-day mission. Units from neighboring Tennessee and Mississippi also took part in the hunt on the assumption that he might have wandered across the state border while trying to avoid a thunderstorm.

"Numerous leads were checked out before CAP Senior Member W. D. Kilpatrick of the Muscle Shoals Squadron spotted the downed plane in a heavily wooded area about ten miles from the Muscle Shoals airport. The doctor had apparently died instantly in the crash."

In 1954, CAP flew 10,671 hours out of a total of 18,260 hours flown by all agencies in domestic air search missions and that year officials of the Air Force Air Rescue Service said that "to maintain the standby force of 12,000 rated second lieutenants represented by the CAP, it would cost the Federal government $46,000,000 a year for base pay and flying pay alone." The cost advantage of CAP in search and rescue was beginning to hit home.

Three years later, Brig. Gen. Thomas DuBose, then commander of ARS, reported that the CAP segment of all air rescue hours flown in 1956 had grown to 77 percent. CAP flew the equivalent of 49 times around the world that year for ARS—12,321 of the 15,797 hours put in by all agencies.

Throughout the rest of the '50s and into the '60s, CAP continued to fly from 65 to 75 percent of the air search and rescue hours recorded by ARS. In just one five-month period between February and June 1961, this amounted to nearly 9,400 hours. The ARS commander reported that CAP "had supported 90 missions and aided in the recovery of 271 survivors." By this point in time, CAP was reporting more than 4,000 light aircraft both CAP-owned and member-owned and some 9,500 licensed pilots.

"As the figures show," he said, "the Civil Air Patrol is our main source of assistance and its contributions make the National SAR Plan an effective tool in the protection of our distressed citizenry."

The magnitude of general aviation operations continued to swell; so did CAP's search and rescue mission. March 1970 was a case in point as the Arizona Wing went on a month-long SAR marathon.

The wing, barely relaxed from nine days of searching for two aircraft the previous month, was first alerted March 8 when a Piper Cherokee was reported overdue on a flight between Bakersfield, California and Phoenix. The Cherokee 140, piloted by Michael Maffei, 30, of Hayward, California, had originally taken off from Oakland, California, and had refueled and taken on two passengers at Bakersfield before departing for Phoenix.

According to FAA officials, Maffei had been briefed concerning bad enroute weather, but had elected to continue his flight.

So at dawn the following day, 13 aircraft from the Arizona Wing, joined by search planes from California CAP units, were launched on the hunt. After two days of searching, the plane was found crashed on

Mt. Baker in the Tehachapi Mountains of California. All three persons aboard were dead and the find was officially credited to the California Wing.

Arizona's fliers, however, were only allowed to stand down for bare minutes. For on the day this first REDCAP closed, a second one was opened. This hunt was for a Cessna 210 missing between Las Vegas, Nevada, and Albuquerque, New Mexico, and flight planned to cross Central Arizona near Prescott.

CAP units from three states, Nevada, Arizona and New Mexico, spent two days aloft in marginal weather when their efforts proved fruitless. Lt. Col. J. B. Gotcher, working as mission coordinator at Deer Valley Airport north of Phoenix, ordered a complete and detailed analysis of the pilot's flying habits and weather at the time of the missing plane's flight.

U.S. Weather Bureau personnel in Phoenix spent two hours backtracking their data before coming up with a weather picture accurate to the nearest foot of altitude, degree of temperature and minute of latitude. This, coupled with information obtained from the pilot's wife ("My husband has a tendency to climb above bad weather. No, there wasn't any oxygen on board the airplane.") led Gotcher to pinpoint the area of Red Lake, north of Kingman, Ariz., as high probability target area.

On the following day, that's exactly where searchers found the wrecked Cessna. The two occupants, both employees of the Atomic Energy Commission, New Mexico, were dead in the wreckage.

Rest for Arizona CAP crews was shortlived, for 16 hours later the wing was again alerted for its third search of the month.

This mission concerned a Cherokee 6 with five persons aboard missing on a comparatively short, 100-mile flight between Safford and Tucson, Arizona.

Severe turbulence, mountain snows and dense underbrush kept mission personnel restricted in their efforts despite days that saw each search aircraft flying as many as three sorties through the rugged peaks of southeastern Arizona.

After 10 days the search was suspended although civilian and sheriff's ground parties remained to comb the area.

Strangely, they were assisted in their work by a prediction from Peter Hurkos, internationally famed psychic.

Hurkos, paid $3,800 as a finder's fee by the wife of the missing pilot, examined a map and articles of clothing in Los Angeles and predicted that the airplane would be found in Taylor Canyon on mighty Mount Graham, southwest of the take-off point.

A day later, a ground party stumbled upon the wreck. It was completely burned and buried beneath trees and invisible even to a police helicopter hovering 20 feet above the treetops. It was in Taylor Canyon on Mount Graham.

But, as one law enforcement officer said at the time, searchers were "90 percent certain that if this plane was anywhere it would be on

Mount Graham and in Taylor Canyon where we did have a sighting . . . so this could have been Mr. Hurkos playing a coincidence."

The five occupants in this airplane also were killed in the crash.

Once more, there was absolutely no rest for the wickedly weary Arizona crews. On March 26, with the Safford search barely closed, CAP aircraft and personnel were routed to Grand Canyon Airport for their fourth mission.

Nightmare for any Arizona search pilot is REDCAP into the Grand Canyon, 14-miles across and 6,000-feet down to its craggy floor. But that's where pilots and observers went in an all out hunt for a Cherokee Six of Grand Canyon Airlines, missing with a full load of six persons aboard between Grand Canyon Airport and Las Vegas, Nev.

The missing pilot, Russ Marsh, 51, of Prescott was a veteran with 15,000 hours. His passengers were all Europeans, members of a German Day School Association touring the United States as guests of the U.S. Department of Health, Education and Welfare. Federal pressure was on. For more than two weeks, until April 12, CAP units from Arizona and Nevada were joined by reserve and active Air Force units from California, Arizona and Oregon, the Arizona Army National Guard, and finally a CAP squadron from California, in scouting close to a quarter million square miles of northwestern Arizona and southern Nevada.

During all of 1970, CAP crews flew 9,383 individual sorties on 369 ARRS directed missions. Hawaii Wing aircraft flew 121 sorties on 76 additional missions and were credited with 12 of the 24 "saves" recorded by CAP that year. In all, CAP was credited with "helping 113 persons in distress" during the 12-month period.

One mission in particular, brought special comment from Brig. Gen. Frank K. Everest Jr., ARRS commander at the time. The mission was a six-wing effort which continued for 15 days when an F-111 swing-wing fighter-bomber was reported missing on a round-robin from Carswell AFB, Tex. In a communication to CAP National Headquarters, General Everest wrote:

"I wish to extend my warmest appreciation for the outstanding performance and participation of the six CAP wings during the recent 15-day search for the F-111. This SAR effort, including all search agencies produced 1,425 sorties and 3,508 hours.

"Of these totals, the Civil Air Patrol flew 81 percent of the sorties and 65 percent of the flying hours. Once again, the Civil Air Patrol has demonstrated its extremely valuable SAR capability not only in professional performance but in significant cost savings to the Air Force."

Official CAP SAR flying hours for 1971 were nearly 31,000 but dropped slightly to 27,400 for 1972. In 1973, the figure was 27,284. The percentage of total flying time recorded by all agencies involved in the ARRS-directed missions remains approximately 80 and with inflation and the fact the dollar is buying less, the savings to the American tax payer becomes more significant each year.

On a typical CAP/USAF SAR mission flown in 1971, the Air Force ran a study which revealed the comparative costs of performing such a mission with CAP volunteers as opposed to costs for currently available USAF aircraft and crews. The result was astounding. According to USAF figures, it averages approximately $400 per hour to operate Air Force aircraft with military crews and support them in the field on a search and rescue effort.

On Air Force-ordered missions, CAP personnel are paid only for the actual fuel and engine lubricants consumed by their aircraft and for commercial communications costs paid out of pocket to complete the mission. The study revealed that this amounts to $11.54 an hour. Over a three year period—1970, 71 and 72—this represented net annual savings to the taxpayer for the 36-months in excess of $30 million. The cost of 78,100 flying hours for CAP was $901,000. The same cost if USAF aircraft had been used would have been $31 million.

Any way you want to figure it, from a practical point of view, from an economic one, or from the point of view of training and experience, the Civil Air Patrol is squarely in the pilot seat when it comes to air search or rescue from New York to California and from Florida to Alaska. But, keeping CAP ready and capable to perform this critical mission is a never ending requirement for training, training and more training.

California MCO, Lou Dartanner, puts it this way:

"The purpose behind search and rescue activities is that of saving lives of those who are lost or missing, while still protecting the lives and equipment of those who are involved in the search. With this in mind, it is essential that all efforts be consistent with the capabilities of CAP emergency services personnel, equipment, facilities, climatic conditions and existing weather. Each emergency situation still presents new and complex problems. This makes it mandatory that once emergency services personnel are selected based on qualification, provided with specialized training and credentialed, they must maintain a high degree of proficiency in their specialty."

One method of aiding CAP personnel to maintain that proficiency is the SARCAP (for Search and Rescue CAP). At least once each 12 months, each USAF/CAP Regional Liaison Officer, in cooperation with ARRS, conducts a simulated search and rescue mission in which all elements of the state's SAR force are exercised. A problem, usually based on a combination of missions, situations found in actual missions, is prepared and presented to CAP to solve. Air Force personnel then monitor the various phases of the mission grading CAP personnel on their performance. In the post-SARCAP critique, no holds are barred and the strengths and weaknesses of the SAR organization are thoroughly analyzed.

USAF evaluators closely scrutinize every phase of the mission from the performance of the Mission Coordinator and the Operations staff through communications and ground safety to flying activities. Flight line crews as well as aircrews are especially checked for ground and flying safety.

"While in the end, it is up to the search crews to find the target," Dartanner points out, "it is the Mission Coordinator who makes the mission tick. He or she not only has to have a handle on the administrative details—it takes a lot of meticulous paperwork for instance, to document the fuel and lubricant bills so that the members get promptly reimbursed—but, also, must have a thorough knowledge of air operations; first to be able to figure out where to deploy the initial forces so they will achieve maximum effectiveness during the early hours of the search when the probability of survivors is highest and, second, to control those forces once they are deployed.

"It takes a lot of aviation savvy and a lot of experience to be a good Mission Coordinator. Some of California Wing MCs have been running search missions for 20 to 25 years. I guess it also takes a special kind of person, one who is virtually unshakable, one who can be hard-nosed when need be and one who can be the soul of diplomacy when discretion demands it. On a mission the MC becomes the operational commander for all accounts and purposes the wing commander for that mission. That's one hell of a lot of responsibility."

The first order of business is to collect every piece of information available about the missing aircraft, from every possible source. Sometimes a considerable amount of good data is available as the result of the FAA communications search, the aircraft flight plan and the preliminary investigation conducted by the Rescue Coordination Center by telephone prior to putting CAP on REDCAP status. More often than not, this initial input is meager and offers little to go on unless the aircraft was on a flight plan and was making regular position reports. Unfortunately, all too few pilots file flight plans in VFR—visual flight rules—weather.

"This is when the MC must also be a good detective," Dartanner explains, "since he must 'build a case' for looking in one direction rather than another. Sometimes hours of painful interrogation of wives and families, business associates, friends, fellow pilots, local FAA authorities at his home airfield, the fixed base operator, his instructor, local law enforcement agencies and even people at other airports where he is found to have visited on this flight or even on previous flights must be conducted. Like a big complex, jigsaw puzzle, these bits and pieces of information are fitted together to provide a picture of the pilot, his flying capabilities and experience, his personal habits and idiosyncrasies. To make the task even more difficult, in most cases where a flight plan has not been filed, we are not alerted to search sometimes for several days and only then after someone has suddenly realized 'old Joe hasn't been around for awhile'. Even with a flight plan on file, it isn't at all unusual to have a pilot who decided to change his route or destination in flight and failed to notify FAA of the change."

Once the aircraft's route has been determined, either from flight plan information or other inputs, the faster, heavier, CAP aircraft are sent on route search. As the search progresses, other search patterns are put into use depending upon the size of area to be covered, the terrain, the weather and the availability of leads which tend to point toward concentrating search capability in specific areas. These include the parallel track pattern, creeping line pattern, the track crawl, the expanding square and the contour search.

The dangerous, difficult task of air search isn't taken lightly by Civil Air Patrol emergency services personnel. Neither are the grisly remains usually found at the end of the search. But, as in all dirty jobs, the people who perform them tend to develop their own colorful phraseology. So it is with CAP search crews, especially the hardened veterans.

In the first place, any old-timer to the search game will tell you, "Don't expect to find anything that resembles an aircraft; most wrecks look like hastily discarded trash." Then, they describe the six basic types of crashes this way:

"Hole-in-the-ground; from steep dives into the ground or from flying straight into steep hillsides or canyon walls. Wreckage is confined to a small circular area around a deep high-walled crater.

"Cork screw or sugar: from uncontrolled spins. Wreckage is considerably broken up and scattered about in a small area. There are curved ground scars around a shallow crater.

"Creaming or smear: from low-level 'buzzing or flat-hatting'; from instrument flight, or attempted crash landing. Wreckage distribution is long and narrow with heavier components farthest away from the initial point of impact.

"The four winds: from mid-air collisions or explosions. Wreckage components are broken up and scattered over a wide area along the flight path.

"Hedge-trimming: where an aircraft strikes a high mountain ridge or obstruction but continues on for a considerable distance before crashing.

"Splash: where aircraft has gone down into water, oil slicks, foam and small bits of floating debris are apparent for a few hours after impact. With time the foam dissipates, the oil slicks spread and streak and the debris becomes widely separated due to action of wind and currents."

What are CAP observers trained to look for? Not airplanes, at least not airplanes that look like airplanes. Instead, they are trained to "see" light colored or shiny objects, smoke, blackened areas, broken tree branches, local discoloration in foliage, fresh bare earth, breaks in cultivated field patterns, oil slicks in water, excessive bubbles in water, discolored water or snow, flashes of sunlight on metal and, of course, personnel and signals where the pilot and passengers have been lucky.

One of those who was lucky and one whose life was saved due to the sharp eyes of an Illinois Wing CAP search team was Navy Lt.(j.g.) J. J. Meder.

Aircraft commander and his observer carefully check the sectional chart prior to departing for their assigned grid during a search and rescue mission. CAP today flys 80 percent of all hours flown on Air Force-directed missions.

Skilled personnel of the regular Air Force like this ARRS helicopter pilot work hand-in-hand with Civil Air Patrol SAR experts when human life is endangered. This CAP senior member points out location of missing aircraft just reported by a CAP search crew to the chopper pilot who will go in to rescue survivors.

An assistant mission coordinator debriefs search crew just returned from sortie where wreckage has been sighted. Precise information on sighting will be relayed via radio to ground teams standing by in area.

CAP cadet checks in a search pilot during a wing SAR evaluation mission. Cadets get thorough training in many ground operations assignments in support of air operations.

CHAPTER 4

"I Came Here To Thank You For My Life!"

October 1954. The huge main ballroom of Washington's Shoreham Hotel. More than 800 Civil Air Patrol members from all over America were gathered for CAP's Congressional Dinner, a bi-annual affair held to report to the Congress of the United States on the achievement for CAP for the previous 12 months.

On the elevated stage at one end of the ballroom, a long head table had been set up. Seated at the table flanking Secretary of the Air Force Harold E. Talbott and Air Force Chief of Staff Gen. Nathan F. Twining, were Maj. Gen. Lucas V. 'Vic' Beau, USAF, the national commander of CAP, Gen. Carl A. 'Tooey' Spaatz, USAF, who had been the Air Force's first chief of staff and later became the first chairman of CAP's National Board, and more than 20 members of the Congress together with ranking CAP officials and other government dignitaries.

General Beau had completed his oral "Report to Congress." Secretary Talbott and General Twining had heartily endorsed the general's report and appropriate members of the Congress had responded. Taking the rostrum again, General Beau said:

"Now, I want to introduce to you an individual who is here because he has something to say to the Civil Air Patrol members present."

All eyes followed the general's to the back of the ballroom where a crewcut young Naval officer on crutches painfully began making his way down the aisle toward the stage. At first, there was a buzz of voices. Then, as it became apparent that each movement was bringing excruciating pain to the young man, the voices were stilled. By the time he made it to the stairs, the room was in complete silence. Courteously, but positively, the officer shrugged away those who moved to help him up the steps. Finally, he reached the podium where he rested for several moments before speaking in a weak but distinct voice.

"I asked to come here tonight," he said, "because I wanted to thank you and all the members of the Civil Air Patrol for my life."

Briefly and sometimes haltingly, Joe Meder took the assemblage back a year to a September day when a CAP search plane found him lying near death in an Illinois soy bean field. When he finished, the room remained silent for 10 to 20 seconds, then as a body, the men and women on the stage and throughout the ballroom rose in tribute to the obvious courage and gallantry of the young officer.

It was several days later sitting propped up on his bed in a Washington hotel room that Meder related all the details of the events which ultimately brought him to CAP's Congressional Dinner. The author was among those privileged to listen.

"You know I'm retired. That's right, retired. Put out to pasture at the grand old age of 25.

"I guess you can't blame the Navy though. There just isn't much you can do with a man after he falls seven miles with only half a parachute to break the fall.

"You know I had heard a lot of stories about the things a man thinks about when he faces death—about how his life passes before him. Maybe I'm different, but the only thing I could seem to think about was who would be driving around in my beautiful new Buick.

"Thanks to the guardian angel of all jet pilots and the dedication of a couple hundred civilian putt-putt pilots of the Civil Air Patrol, I'm driving the Buick myself and I can even limp around on shanks' mare with the help of a cane. They tell me I might even be able to throw away the cane in a few years after my 27 fractures heal completely.

"It all began on a cool September morning about a year ago. The old man called me in and told me he wanted me to give a lecture on Naval Justice to a group of Air Reservists at Minneapolis. I was based at Oceana Naval Air Station on the Virginia coast and as long as my element leader, Jim Graham, and I both needed some cross country navigation practice the commander told us to take our Banshees and make the trip do double duty as a dead reckoning training mission. The Navy figures its jet jockeys never got enough training and Fighter Squadron 82 was no exception.

"Jim and I spent the next morning getting things in shape for the trip and checking with our plane captains to be sure the aircrafts were in the pink. Shortly after lunch, we lit up and shoved off.

"The flight to Glenview Naval Air Station near Chicago was uneventful. It was a good day. I mentally gave myself a pat on the back as I greased the jet on the runway at Glenview and rolled up to the refueling line. Here I was fat and sassy—a firecan jockey with more than 800 hours, 300 in jets and a whole skin. It never occurred to me that in a few hours I wouldn't want to bet my chances of living against a plugged dime.

"At Glenview, we took on a fresh load of 145 octane gasoline. The Navy used gasoline instead of kerosene because on carriers you haven't enough storage space to carry two kinds of fuel. I filed our flight plan to Minneapolis as leader and we trotted out to the planes planning to be in Wisconsin in time for an early steak dinner and then 'what have you.'

"During the preflight inspection, I found a leak in the main fuel manifold and asked Jim to take a look. He was the squadron Maintenance officer and he knew the McDonnell Banshee from nose to tail like he knew the palm of his hand.

"One look and he turned to me, 'Why it didn't blow up on the trip up, I don't know!'

"The sprung manifold had let about two gallons of high octane into the belly of the plane where it sloshed around a few inches from the red hot tail pipe.

"Jim personally repaired the leaky manifold and pronounced the plane ready to go. I want to make it clear right now, that I'm dead sure that manifold had nothing to do with what was to come later.

Meder changed his position on the bed and continued:

"By the time we were ready to go, our clearance had expired. In operations we learned that the weather was the same as had been reported earlier—ceiling and visibility unlimited at our destination with a little scattered cloud formation at 12,000 feet. The aerology officer reported a squall line between Glenview and Minneapolis, but he didn't consider it serious. We filed for 5,000 feet IFR through the front.

"The clock on the instrument panel indicated 1910 as I fire-walled the throttle and the acceleration shoved me back against the seat cushion like a giant hand. A Banshee really gets away like a scared cat when you pour on the coal.

"It was beginning to haze up as we made our rendezvous over Libertyville. We were supposed to practice dead reckoning so I took the lead with my radio off while Jim monitored the radio ranges with the birddog. If I got off the track and couldn't seem to get my bearings, Jim would take over and get us to our destination.

"My dead reckoning had been good from Oceana to Glenview. We flew at 40,000 feet and I was only two miles off course at the end of the line. I hoped it would be that good as we navigated around the heavy squall line which began making itself evident ahead.

"We cooked on. It had been 15 minutes since rendezvous on a course of 305 degrees. The front which the aerologist said wasn't serious looked mighty menacing now. The tops of the nasty clouds towered above us—way up.

"I knew now we couldn't continue on DR. I couldn't see the check points. A flip of the radio switch and we were in business again. I called Jim and told him I was turning on my ADF. 'What's a little weather,' he joked, 'we can always smell our way to a Navy base.'

"The optimism lasted only five minutes. Without so much as a fare you well, my ADF quit cold. I switched to the UHF and called Jim. 'My ADF is gone,' I told him, 'you'll have to take over!'

"He gave me a 'Roger' and slid up past me. The front now looked like a huge frowning wall of boiling clouds. There was no percentage in trying to go through. Sometimes the turbulence in those big, flat-headed nimbus clouds will tear a plane to pieces. It almost happened to me once before in a Panther. I certainly didn't want to go through it again.

"We held a quick conference and decided to head south parallel to the front climbing until we could go over it. The huge puff balls began to drop beneath us at 36,000 feet but we continued on for another 2,000 to give us plenty of clearance. After leveling off, we kept on 100 percent power to accelerate to cruising speed—about 480 knots.

"I called Jim and said I was sliding over to his right wing. I was just about in position when the Banshee was jolted to a sharp explosion. The loud bang was followed by a sound like a handful of metal being thrown into a meat grinder.

"The ship began to yaw with a loss of power in the left engine followed by a hissing sound like a high pressure fuel line rupturing. My eyes scanned the instruments. Port oil pressure—ZERO! Port tachometer—unwinding! I began falling behind. Shut down the engine, that was the thing to do, before it tore itself out of the fuselage. I reached for the fuel shut-off valve and called Jim. Got him on the fifth try. 'I'm in trouble. I think I've got a MAYDAY!'

"Jim looked back. 'For Christ's sake, Joe, you're a ball of fire!' he screamed in the RT.

"I cut the throttle and popped the speed brakes. A quick look over my shoulder revealed a stream of flame and molten metal streaming back past the tail.

"I called Jim back telling him I was bailing out, disconnected my oxygen hose and radio cords and pulled the face curtain. The last look I had at the instrument panel showed the altimeter reading 36,100 feet—seven miles straight up.

"The ejection worked fine. Pulling the face curtain fired the cartridge blowing the canopy away and firing me out of the cockpit like a piece of toast out of a toaster. Everything worked just the way it was supposed to. When I released the curtain, the wind caught my eye shield snapping the helmet against the back of my neck. Otherwise it was strictly routine.

"I was still on top. The sky was clear as a bell above me and there were stars all over the place. I pulled the quick release on the seat belt, but the seat stuck to me like it was glued. I shoved it away and let myself fall awhile. I was just about ready to pull the rip cord when something bumped me in the butt. It was the seat again. It had floated back. I experienced an eerie sensation like the seat and I were suspended along in space. There was no feeling of movement.

"I rolled over and kicked the seat away reaching for the rip cord as I did so. That was the closest I came to cracking up during the entire horrible adventure. The rip cord wasn't under my left armpit where it should have been. Momentarily crazy with animal fear, I clawed frenziedly at my side where the D-ring should have been.

"It was only when I had clawed through my flying suit, the T-shirt under it and felt the pain of my nails tearing into the flesh over my ribs that I got hold of myself.

"Use your head, Joe, I told myself. The D-ring couldn't get away. It is fastened to the rip cord and can't be pulled free without the chute opening.

"I began methodically following the flexible cable from the back pack, over my shoulder and down to the welcome solidity of that beautiful, shiny steel ring. It had pulled from its retainer during the ejection. I found it near my knees and yanked it for all I was worth.

"The chute opened without any excessive jerk, but I paid little attention to it then. I was in the center of the storm. Hail as big as golf balls bounced off my helmet. Lightning flashed at arm's length. Sometimes I felt like I was going up instead of down as I watched the clouds boil around me like a cauldron of hot steam.

"Aside from being tossed around by the storm, the descent felt normal until I broke out at 8,000 and saw I was spinning. When I reached up for the shroud lines to stop the spin, I saw why the opening hadn't jolted me. Four complete panels of life-saving nylon were missing from hem to top and another was ripping.

"Calmly, without emotion, I thought I've had it for sure. That is when the thoughts of my new Buick seemed to wipe out the horror of the moment. It seemed to steady me and I took stock of my situation.

"Visibility was good. I could see towns and automobiles moving along the roads. As I neared the ground, I realized the wind was blowing and I had a ground speed of about 55 knots. I guess I knew I didn't have much chance of surviving the blow, but somehow with that fifth panel ripping I was reaching for old mother earth with my toes.

"There was no sense of shock or pain with the impact. I guess it was because the wind was knocked out of me and I passed out for awhile. When I came to, I was in a bean field. I could hear automobiles about a quarter of a mile away but there was high ground in between me and the road. Still no feeling of pain. I stood up to make for high ground. I had gone a few steps when I felt the distinct sensation of walking in mud up to my hips. I looked down at my legs. That's when I got a real jolt. My right ankle was turned in a right angle to the leg. My left ankle was turned out. I was walking on the stumps of my legs.

"The situation didn't hit me at first. I couldn't seem to comprehend it. I continued on. About 40 feet from my parachute I felt a warmness in my chest and also began to feel faint. I knew then I had better get back to the chute. I had a better chance of being seen when morning came."

Again Meder stopped, squirmed in discomfort, and continued:

"Down on my hands and knees, I began to drag myself along. When I reached the chute, I realized I would have to try and give myself some first aid. First, something to deaden the pain. I was in for another bitter disappointment. Not even aspirin. There was nothing to do but go ahead and try to set the ankles before the pain got too bad. The right ankle popped right into place with the first try. I bound it with parachute shroud lines. It turned out to be set perfectly and the medics never had to touch it again. The left ankle was a real mess and the best I could do was to wrap it in shroud lines.

"The rain still poured down. I inflated my rubber life raft. Placing the chute between me and the cold ground I pulled the raft over me for shelter. As snug as I could be under the circumstances, I reached for a cigarette. One drag and it went out.

"Thinking it must have been wet, I got out another one. Again, one drag and it went out. I tried a third. The same thing happened. I found my flashlight to see if I could find one dry smoke in the pack. There was another shock in store for me. The three I had lighted were dripping scarlet. Blood pouring from my mouth had put them out as fast as I lighted them.

"Still somehow without emotion, I crumpled the pack and threw it away, thinking to myself, I won't need them again!

"During that night I was alternately awake and unconscious. Once a small plane flew over in the dark and then flew back again. I tried to light a night flare. They wouldn't work. Shortly after dawn, I saw a man on a tractor apparently cutting grass along a road some distance away. I yelled myself hoarse and shot off a smoke flare. He didn't see it or hear me. I knew I couldn't last much longer. My only salvation seemed to be in making the road about 400 yards away. I carried one smoke flare with me and began dragging my useless body along in the mud.

"During the next four hours, I made about 50 yards before giving up in complete exhaustion. It must have been an hour later when I heard an airplane engine and a little yellow plane came hedge-hopping over the trees not more than 150 feet above the ground.

"Thank God, he saw me. He circled and gunned the engine to signal me and then lit off again over the trees. I passed out with the burning flare in my hand. Luckily it stopped where it was supposed to or I would have had a badly burned hand to add to my other injuries.

"The next thing I knew a man in an Air Force uniform was running across the mud field to me. As he came up to my side, and knelt I struggled to a half sitting position.

"I remember he wore the silver oak leaves of a lieutenant colonel.

"I tried to get out in the open where you could see me, I gasped and then passed out again.

"The next few hours, in fact, the next few days, were a nightmare of pain sprinkled with moments of consciousness when I woke up only to get another shot or take another pill to put me back to sleep.

"When I was able to sit up and take nourishment, as the old saying goes, my doctor, Lt. Cdr. Francis McCullough of Great Lakes Naval Hospital and one of the greatest orthopedic specialists in the world for my money, told me I had been pretty heavy on the 'no—go' side when they got me to the hospital. If it hadn't been for that little yellow plane, I certainly wouldn't have lived much longer, he informed me with great emphasis.

"When he figured up the box score, I had 14 fractures of the left ankle, one of the right ankle, three compression fractures of the back, six broken ribs—two puncturing the pleural cavity, a pelvis cracked completely across and a lung filled with blood.

"I was especially anxious to thank the men who were responsible for my rescue and a few days later they brought Lt. Col. Loren Whan, director of Communications for the Illinois Wing of the Civil Air Patrol, to see me. I learned that Whan had been the first man to reach me and also had directed the huge, state-wide search that saved my life. He told me that an hour after my wingman reported the bail-out and my Banshee drilled a 75-foot crater into the ground near Hooppole, Ill., CAP members from all over Illinois and Indiana were on their way. During the hours of darkness, they led a ground search aided by police, American Legionnaires and other volunteers over a 40-square-mile area.

"Promptly at dawn their lightplanes took off on the air search. He told me two of his pilots, Vince Causmaker and John Zonge, actually found me. One piloted the little yellow plane I saw before passing out again. The other spotted me during the period before Whan led the ground rescue crew to my muddy bean field.

"Now that the Navy has decided I should play the role of a has-been, I find I miss the smell of burnt gasoline, the roar of jet exhaust and the thrill of torching around the sky. I'm not through with aviation, however. They tell me I can still qualify to fly the little putt-putts and I'm looking forward to getting into the blue again soon.

"Needless to say, I have joined the Civil Air Patrol myself as a member of the New York Wing.

"Who knows, maybe someday I'll get the chance to even the score and save the life of another airman who falls out of the sky. I hope so."

Meder's story and his visit to Washington where the Civil Air Patrol was reporting to the Congress is most appropriate to this historical overview of CAP's first 30-odd years of public service for it underlies the unique status of this nonprofit benevolent organization—one of less than 60 holding a charter granted by the Congress and the only one also established by law as the civilian auxiliary of the United States Air Force.

CHAPTER 5

CAP And The Air Force

The year 1946 was a year of trauma and a year of triumph for the men and women who during second world war had earned the name "Flying Minutemen" the hard way with five long years of service and sacrifice.

During the preceding 12 months, as victory first in Europe and then in the Pacific became a reality, they had seen their mission gradually whittled down to virtually nothing. Anti-submarine patrol, aerial target towing and border patrol had long since been terminated. Courier service was waning rapidly. New emphasis was placed on the recruitment and training of young men and women to take their place in the huge, commercial aviation industry then on the horizon but, while many CAP senior members applied themselves industriously to this activity, it seemed a pale and lackluster substitute for the high adventure and drama of the operational missions on which most of CAP's wartime members thrived.

To get more action back into the organization its leaders began a massive airport marking program in support of the CAA. The intent was to provide an identification easily recognizable from the air for "every town and hamlet" in America. A lot of towns and hamlets and many large cities soon were marked conspicuously with huge, yellow letters spelling out the name with an arrow pointing to the nearest airfield.

In other areas, CAP local units organized aviation breakfasts, local area air tours, and fly-ins. With the approval of their national headquarters, CAP units undertook to sponsor air shows and exhibitions—the purpose to get private flying, which had been discontinued during the war years, back in motion.

Still it wasn't enough. Strong squadrons floundered and just managed to keep their programs intact. Weak ones failed and dropped out of sight. Morale already at an all-time low took another blow when Stearman PT-17 aircraft loaned by the Army Air Forces to CAP were withdrawn shortly after V-J Day. Then early in 1946 the other shoe dropped—CAP was notified by the military that all direct financial support for their program would cease on March 31.

It didn't help that Col. Earle Johnson (later promoted to Brigadier General posthumously) who had been CAP's wartime commander was placed in temporary duty elsewhere leaving the reins to Col. Harry Blee, his deputy. Blee subsequently named "acting National Commander" addressed his full energies to keeping CAP's head above water but many wing commanders and other CAP leaders took the entire episode as a "breach of faith" on the part of the AAF and some took their dissatisfaction to their congressmen.

The first three months of 1946 saw marathon meetings in Washington between Civil Air Patrol leaders including wing commanders from all across the nation and the top echelon of the Army Air Forces. These meetings were highlighted on March 1 by a "Congressional Dinner" hosted by the Civil Air Patrol at which President Harry S. Truman, members of the Congress and Gen. H. H. "Hap" Arnold, wartime commanding general of the AAF, were honored guests. Some accounts of this affair describe it as a "thank you" gesture on the part of CAP for the wartime support given it. To an extent this certainly was true. But, a number of old-timers candidly point out that it more rightly could be described as an exercise in political acumen. In retrospect it can be said that it certainly proved to be the right thing to do at exactly the right time.

It was well known that President Truman had long been an admirer of CAP and its contributions to the nation. The current Army Air Forces commander, Gen. Carl A. "Tooey" Spaatz, made no secret of the fact he fully supported the CAP and saw a peacetime mission for it. What was needed was the unequivocal support of the Congress to put into motion the future plans for CAP being formulated at the crucial meetings.

That public display of solidarity between the White House, the leadership of the Army Air Forces and the Civil Air Patrol had the desired effect. Later that spring the 79th Congress enacted Public Law 476 granting CAP a charter to operate as a non-profit, benevolent corporation. For the first time, CAP had a foundation established in statute (throughout World War II it existed only by virtue of executive order). Under the law, the new corporation was entitled to a tax-exempt status but it remained for CAP leaders, headed by Col. Paul W. Turner, to work long and hard to bring about the ultimate amendments to the Internal Revenue Code which made the exemption a reality. Under the code, CAP members are allowed a deduction for their out-of-pocket expenses (uniforms, special equipment required to perform their missions, mileage and similar costs resulting from participation in official CAP activities). Also, individuals and corporations may deduct donations made to CAP. With CAP an official, non-profit corporation, the military also would find it easier to provide certain types of limited support. The aggregate effect was a real improvement in the status of CAP and on the basis of its new status the corporation embarked on a man-sized program embracing 9 principle points:

. inform the general public about aviation and its impacts;

. provide its seniors and cadets with aviation education and ground and pre-flight training;

. provide air service under emergency conditions;

. establish a radio network covering all parts of the United States for both training and emergency use;

. encourage the establishment of flying clubs for its membership;

. provide selected cadets a two-week encampment at air bases;
. encourage model airplane building and flying;
. assist veterans to find employment; and
. contribute services to special projects such as airport development, the survey and marking of emergency landing areas and the survey of dangerous flying areas in mountainous regions.

In addition, the newly chartered CAP corporation was prepared to take on certain tasks assigned by the Army Air Forces even though there was no official basis for the service to respond in kind. Although there were plans to correct this situation in the mill, little could be done at this time since the status of America's military air arm also was in a state of flux. The men who had most recently demonstrated that air power could be the deciding factor in a major international conflict were pushing strongly for an independent air force. Unfortunately the path was strewn with stumbling blocks placed there—both overtly and covertly— by non-flying Army leaders who did not want to lose this vital new mission and by both flying and non-flying Naval types who felt that land-based air would represent less of a challenge in the competition for roles and missions if it continued to be subservient to the Army. It was a time of turmoil for the new service striving to be born—a period when little could be done to advance CAP hopes for official status as a military auxiliary.

On September 18, 1947, a simple ceremony attended by Army Secretary Kenneth Royall, Defense Secretary James Forrestal and Navy Secretary John Sullivan took place in Washington. The principles were Chief Justice of the Supreme Court Fred Vinson and W. Stuart Symington, a long-time booster of air power. When the ceremony was concluded Symington was the First Secretary of the Air Force. Simultaneously the United States gained a third independent military service, one devoted exclusively to building and maintaining air power second to none.

This climaxed the series of events which had begun on July 26 when Congress passed the National Security Act of 1947 creating the National Military Establishment as an executive department headed by a civilian Secretary of Defense and establishing three equal and independent departments—the Army, the Navy and the Air Force. The dream of military flying men for three decades then came one more step closer to reality as President Truman signed Executive Order 9877 which prescribed the roles and functions of the three services. But it took the September 18 ceremony to give the U.S. Air Force a "birth certificate."

The man selected to become the first chief of staff of the Air Force was the same Tooey Spaatz who, as a major in 1929, with Capt. Ira C. Eaker and Lt. Elwood "Pete" Quesada, circled San Diego for 151 hours in a lumbering Fokker trimotor setting a world endurance record; the man who set the Eighth Air Force on its victory course over Europe; and the man who, in the latter days of the war, commanded the Army Air Forces. This was the same man who, at that

1946 CAP Congressional Dinner drew the analogy between CAP and volunteer fire brigades of an earlier era.

"There also was a time, not long ago," Spaatz told the assemblage, "when America was in danger of something worse than fire. The Air Force was not prepared to meet that danger, not equipped for adequate defense of the country, much less for offense overseas. It was then, in 1941, that the Civil Air Patrol was founded somewhat as a firebucket project. That too was the volunteer spirit . . .

"We must pass on our air experience—not only in the Air Forces, but in every section of the country . . . sparking the advance will be the Civil Air Patrol."

Notwithstanding the back-breaking task facing him in establishing the organization and structure of the new Air Force, Spaatz found time to stay in touch with CAP leadership and personally backed the move to give the Civil Air Patrol official status in the Air Force family. Essentially, Spaatz had a mandate to do so. To understand the situation it is necessary to look at the Air Force objectives as set out on October 27, 1947, at the time the USAF accepted CAP from the USAAF. One of the objectives was to press for legislation to legalize the provisions necessary for the Air Force to provide assistance to the CAP. This culminated in a request for legislation that ultimately would:

(1) create CAP as the official auxiliary of the USAF;

(2) direct the USAF to support the Civil Air Patrol in accomplishment of its objectives; and

(3) authorize the Secretary of the Air Force to utilize the resources of the CAP.

That move bore fruit on May 26, 1948 when the Congress enacted a second piece of legislation bearing on the CAP— Public Law 557, 80th Congress, Second Session—making CAP the auxiliary of the new United States Air Force.

With the exception of Earle Johnson who served as CAP national commander from March 1942 until February 1947, the organization did not enjoy the services of what could be rightly termed a "full time" commander until late in 1947. Maj. Gen. John F. Curry—December 1941 to March 1942; Maj. Gen. Frederic H. Smith, Jr.—February 1947 to October 1947 could, in retrospect, almost be described "caretaker" commanders. It is true each contributed to the mission and to the growth of the Civil Air Patrol, but both had short tenures and both were concerned with myriad ancillary duties.

If the development of the Civil Air Patrol as we know it today can be attributed to any individual, two men must share credit—wiley, perspicacious Tooey Spaatz and the man Spaatz selected to ramrod the "new" Civil Air Patrol envisioned in the pair of public laws enacted by the Congress—Maj. Gen. Lucas V. "Vic" Beau.

Vic Beau was an airman's airman. He came from a flying family; his father operated one of the pioneer flying fields at Mineola, Long Island, and as a boy young Vic would trolley daily from his Larchmont home to soak up the aroma of gasoline and castor oil. Aviation occupied more and more of his thoughts as he acquired a formal education at Mamaroneck High School, Syracuse University and Cornell University.

He joined the New York National Guard and at the outbreak of World War I, was ordered to active duty. It was logical that in the military he should get his sights on the air, and he quickly ended up as a flying cadet in the Aviation Section of the Army Signal Corps.

The American Jenny, the French Spad and Neiuport and the British Sopwith Camel became old friends as Vic served as an instructor pilot at the great allied advanced flying school at Issodoun, France. And, after the war, Beau continued his military career expanding his cockpit familiarity to the B-17 Fortress and B-24 Liberators of World War II and developing a reputation as a "can-do" commander with the rare capability of combining a hard-assed determination to get things done and done the right way with a natural penchant for diplomacy. This was exactly what the job needed. To whip the postwar CAP into shape would require the patience of Job, the tenacity of a bulldog and the flair of an international diplomat.

Vic Beau took up the challenge with a vengeance in October 1947 and held the job for eight long, productive years. For all of that period the team of Spaatz and Beau remained intact in one form or another. When Spaatz retired from active Air Force duty in 1948 he became the first chairman of the National Board of CAP, a position he retained until 1959. Vic Beau had the ear of Tooey Spaatz during his tenure as USAF Chief of Staff insuring CAP got the high level Air Force support it deserved and when Tooey became chairman of the board he maintained CAP's front office presence with the Air Force through his close relationship with his hand-picked successor as chief of staff, young Gen. Hoyt S. Vandenberg, and later with Vandenberg's successor, Gen. Nathan F. Twining.

Recently, Beau dropped into Washington's Army-Navy Club and reminisced briefly about his early days as CAP National Commander.

"You know," he said, "it was no picnic. I got a bucket of worms but Tooey had warned me and I was ready for it. We knew how important the Civil Air Patrol could be to the Air Force and the country if it could be made to jell and that's what Tooey hired me to do—get off my butt and get the job done.

"From the outset I saw it wasn't going to be any office job. Just from what I could see from Washington, I knew we had some damned good wing commanders and some that were punk. We had some great squadrons and some that were lousy. The real problem was to slice the bad away so the good—and most of it was good—could thrive and grow.

"There was nothing to do but hit the road. I think I traveled the better part of those first two years hitting wings, groups and squadrons all over the country. They told me I showed up at the darndest times. The most important thing to CAP is good commanders. With the help of local aviation enthusiasts, state aviation officials and the congressmen from the various states we located the right men for the wing command jobs and then persuaded them to take on the responsibility. We didn't make all the right decisions the first time around. In a couple of instances it didn't take. There was nothing to do but make a change. It wasn't too long, however, before we had a bunch of tigers and CAP really began to take off."

The spirit and the verve implicit in the choice of words and the manner of speech Vic Beau assumes today when he recalls his tenure as National Commander go a long way toward explaining his resounding success in the job. He gave it everything he had and with Vic Beau, that is plenty.

It was under General Beau's leadership that the close working relationship between the active establishment and the CAP first was forged, then tempered; the eminently successful International Air Cadet Exchange was founded; the solidification of a lasting relationship between CAP and the Federal Civil Defense Agency (now the Defense Civil Preparedness Agency) was accomplished; the cadet summer encampment program was expanded; permanent Air Force-CAP liaison offices staffed with active duty Air Force professionals were established at each wing and later at the eight regions; a close relationship between the CAP and the new Air Force Academy was established; military lightplanes uniquely suited to CAP's mission (surplus to the needs of the active services) first were loaned to CAP then were turned over outright to auxiliary; vehicles and communications equipment also excess to the needs of the Department of Defense were made available; the CAP's formal aerospace education program was launched; and the Air Force's little brother became, in fact, "the bluesuit" presence in more than 2,000 communities across the nation where Air Force representation otherwise would be lacking.

Obviously, Tooey Spaatz and Beau didn't accomplish the job alone. They had the help of several hundred officers, airmen and Air Force civilian employees assigned over those early years to assist in the job. But, even more important, they had the confidence and wholehearted support of scores of dedicated CAP leaders—Col. D. Harold Byrd of Texas, Col. Harry Coffey of Washington, Col. Alfred Waddell of Tennessee, Col. Hank Zoller of Kansas, Col. Orlando Antonsanti of Puerto Rico, Col. Warren Reimers of Mississippi, Col. Joseph L. Floyd of South Dakota, Col. Howard Freeman of California, Col. Allen C. Perkinson of Virginia, Col. J. Gibbs Spring of New Mexico, Col. Louisa S. Morse of Delaware, Col. Charles Boettcher of Colorado, Col. Benjamin F. Dillingham of Hawaii—a seemingly endless list of men and women dedicated to working diligently for their community, state and nation, and for air power.

On the retirement of General Beau, the Secretary of the Air Force named Maj. Gen. Walter R. Agee as national commander. General Agee carried on in the manner of Vic Beau through his tour of duty which ended with retirement in March 1959 when Brig. Gen. Stephen

D. McElroy assumed command.

A young general, a command pilot, known as a "charger", McElroy came to CAP from an assignment as chief of staff at the Air Force Academy. He had the opportunity to see first hand the payoff of the CAP cadet program—stalwart, motivated, Air Force-minded young men flowing from CAP into the academy. He also had a good idea of what needed to be done next. Under the leadership of Spaatz, Beau and Agee, CAP had matured. It was healthy, robust and well coordinated but like young athletes often get, a little over-organized. McElroy had the right prescription—streamline the Air Force side of the house and counsel the corporate leadership to do likewise. CAP took McElroy's counsel to heart and entered the 1960s with its sights on even loftier goals than those already reached.

A number of developments transpired during the 1960s, however, which combined to slow the Civil Air Patrol's dramatic growth pattern. In fact, the first of these occurred early in 1959.

Traditionally, CAP had been headquartered in the nation's capital, a part of that pulsing nerve center that is the seat of government. Thus, not only did the organization and its appointed commanders enjoy close personal relationships with senior members of the Air Staff, the very location of the CAP headquarters made it a matter of a 30-minute, cross-town jaunt to personally coordinate a program, brief on an accomplishment or justify a request. Based at Bolling AFB, Hq., CAP-USAF (the Air Force headquarters for CAP) administratively reported to Headquarters Command, USAF (but, for all accounts and purposes was operationally directly to Hq. USAF.)

Not only was CAP National Headquarters now moved physically from its proximity to the Air Force decision-making level (to Ellington AFB, Texas, 1,500 miles away) but in the chain of command it was placed under the Continental Air Command both administratively and operationally. This, in the real world of intra-Air Force relationships constituted a significant demotion in terms of the stature and prestige (not to mention just plain convenience) inherent in an organization's place in the military pecking order.

During the early 1960s other forces were at work in the military, in the nation and in the world that would tend to nudge the Civil Air Patrol even further from the favored position it occupied in the Air Force hierarchy during its first 20 years. The cold war was heating up. More and more U.S. manpower and financial resources were going into Southeast Asia. The nation had embarked on a $20 billion space program with a goal of putting man on the moon before the end of the decade and unbeknownst to the public another $20 billion was being earmarked by the Department of Defense for the development and deployment of sophisticated military space systems that were critical to our national survival.

A lot of priorities had been shifted and a lot of new ones were emerging. For many years, military personnal assigned to CAP had been virtually hand-picked with considerable emphasis on those special qualities necessary for an individual to effectively perform the oft times touchy task of liaison between a civilian organization and the active military establishment. Now, like many other elements of the Air Force that were not combat-oriented, CAP was considered "soft-core" and personnel requirements were met through the "pipeline."

When General McElroy's talents were required for another assignment late in 1961, his deputy Col. Paul Ashworth assumed command of CAP and continued in that capacity until August 1964 when he was succeeded by Col. Joe L. Mason. Mason was the national commander until May 1967.

Both Ashworth and Mason applied themselves assiduously to the task and both earned the gratitude of the leadership as well as the rank and file of the CAP corporation but in the stark light of reality it just wasn't the same.

The "military mystique" dictates that the degree of importance placed on a command, the amount of attention it receives in the constant battle for funds, resources and personnel can often be measured in direct proportion to the rank of the commander assigned. Traditionally, the Civil Air Patrol had rated a general officer. Now, in the eyes of most of the Air Force, that berth had been "downgraded", an indication that the CAP had been moved to the back burner.

It is important to point out that despite a general lessening of morale throughout the organization during the major part of the 1960s, this did not deter the Civil Air Patrol from doing its job. Its operational missions continued to grow and the membership—the pilots, observers, mechanics, radio operations, ground personnel—continued to more than meet the challenge. It also learned to forage a little more for itself and to depend a little less for support on the active military establishment. Many wings soon began taking special pride in this new spirit of self-sufficiency.

At the same time, many influential veteran CAP leaders began making it known directly and through their Congressional channels that they did not appreciate what they considered a reduction in stature, particularly in view of the ever increasing level of mission activity—particularly air search and rescue—CAP was accomplishing. Somewhere along the line the word got through and in May 1967, the Civil Air Patrol got back its "star status" with the appointment of Brig. Gen. William Wilcox as National Commander. Wilcox remained in the position only a year. He was succeeded by Maj. Gen. Walter B. Putnam who also served only a one-year tour. But between them, Wilcox and Putnam managed to see that CAP recovered much of its lost prestige within the Air Force establishment. During the same period, a pair of organizational moves also aided measurably—CAP got a new, permanent headquarters at Maxwell AFB, Ala., home of the Air University, and the Continental Air Command was abolished. With CONAC's demise, Hq., CAP-USAF again was placed under Headquarters command reestablishing its direct access to the Air Force front office with promise of a full tour.

Brig. Gen. Richard N. Ellis became National Commander in November 1969 and for the first time in a decade a sense of real stability began to permeate the organization. Under Ellis's direction, the Civil Air Patrol again began to move ahead. There was a resurgence of the spirit that had marked the post-World War II period. New programs were instituted. Old programs were rejuvenated. CAP turned the corner into its third decade of public service again setting its sights on new and even more ambitious goals, and the 1970s showed definite signs of treating the organization considerably better than the decade before.

The current National Commander of the Civil Air Patrol (since November 1972) is 54-year-old Brig. Gen. Leslie J. Westberg, a tall, quietly forceful, officer who flew in combat not only in World War II, but also over Korea and in Vietnam. He is a jet-age general having flown 240 RF4C tactical reconnaissance missions in Southeast Asia before being named Chief of Staff of the Seventh Air Force at Saigon.

In many ways, Leslie Westberg is very much like Vic Beau at least with regard to the value he places on the Civil Air Patrol as an auxiliary of the Air Force. Like Beau, he feels that one of CAP's most valuable contributions is that it gives the Air Force—and thus the maintenance of air supremacy—a presence at the grass roots, the community level across America.

Westberg, a patriot, dedicated to preserving the American way of life, feels strongly that only with public understanding and support can his service, the Air Force, accomplish its role in effecting that preservation. He is convinced that the Civil Air Patrol can be a powerful instrument for enhancing the required understanding and insuring that support. Westberg also believes that the level to which the Civil Air Patrol can be expected to support and assist the Air Force is in direct proportion to the degree to which the Air Force understands and supports the CAP.

In charting the relationship in this book between the Civil Air Patrol and the Air Force no little emphasis has been placed on the men who have been assigned the task of making that relationship work. The nuts and bolts of that relationship are clearly called out in law and in regulation but it is the men who must interpret them. It is necessary, however, to fully understand the scope of the statutes and regulations and the limitations inherent therein.

In Public Law 476 which gave CAP its congressional character, the purposes of the Civil Air Patrol are broadly stipulated in this manner:

"(a) to provide an organization to encourage and aid American citizens in the contribution of their efforts, services and resources in the development of aviation and in the maintenance of air supremacy, and to encourage and develop by example the voluntary contribution of private citizens to the public welfare;

"(b) to provide aviation education and training especially to its senior and cadet members; to encourage and foster civil aviation in local communities and to provide an organization of private citizens with adequate facilities to assist in meeting local and national emergencies."

The Civil Air Patrol and the United States Air Force maintain a civilian-military relationship based upon the Civil Air Patrol's status as a USAF Auxiliary under Public Law 557. The law, and its amendments, did not change the character of CAP as a private corporation, nor make it an agency of the U.S. Government, but gave the Secretary of the Air Force certain authority to furnish assistance to CAP and to accept and utilize the services of CAP in the fulfillment of the noncombatant mission of the Air Force. As such, the CAP's services to the nation and the USAF are: (1) voluntary, (2) benevolent, and (3) noncombatant. These services may be employed both in times of peace and war.

It is the Air Force's responsibility to provide technical information and advice to those CAP members who organize and train other CAP personnel; who develop CAP resources; and who use CAP personnel and resources. In addition, the Air Force makes available certain services and facilities required by CAP. Such assistance, however, is restricted to specific areas by acts of Congress, and cannot interfere with the performance of the Air Force mission.

The Air Force makes available CAP aircraft, motor vehicles, communications equipment, spare parts, rescue and safety equipment, and office equipment surplus to the needs of defense. Based on the availability of aircraft, flight crews and travel funds, the Air Force provides airlift services for various CAP programs such as summer encampments, aerospace education workshops, etc. In addition to airlift, the USAF takes CAP cadets on orientation flights to supplement their academic training.

Also subject to availability, Air Force base commanders provide meeting places and classrooms for local CAP units and parking spaces for CAP aircraft, and furnish guidance and training literature to enrich the CAP program.

Cadets attend summer encampments held at Air Force bases throughout the nation. The Air Force actively supports this training function not only furnishing quarters and office space but also providing advisors, instructors, and training aids. While the cadets are on encampment, they are provided emergency medical care. Cadets eat at Air Force dining halls. In addition to base cafeteria privileges, cadets have access to the base theater, bowling alleys, swimming pools, and airmen's clubs.

The Air Force encourages its reserve components to contribute their services to the CAP. Inactive reservists earn credit toward retirement by serving in a variety of capacities, including instructors and advisors to CAP cadets and instructors in the senior training program. Civil Air Patrol's aerospace education program also is helped greatly by those Air Force reservists who are professional educators. They serve as consultants, assistants and guest lecturers at CAP aerospace education workshops held at colleges and universities throughout the nation.

CAP is a civilian corporation made up of volunteers who pay dues for the privilege of being a member and rendering a service to the nation. Although CAP members wear an adaptation of the Air Force uniform, have an organization that is patterned after that of the Air Force and perform their duties in a military manner, they still are civilians.

The CAP is organized into eight geographic regions. The eight regions are subdivided by the states which fall within their boundaries, and each state is classified as a wing—giving a total of 52 including the District of Columbia and the Commonwealth of Puerto Rico. Each wing is subdivided into sectors (where authorized), groups (optional), squadrons and flights, according to the organizational needs.

The highest governing body of the Civil Air Patrol Corporation is the National Board, headed by an elected chairman. The only non-civilian on the National Board is the National Commander, who is a member of the Civil Air Patrol Corporation and also an active duty Air Force Officer. All other members of the National Board hold CAP grade and include the eight CAP regional commanders, the 52 CAP wing commanders, plus the National Finance and National Legal Officer.

At least once annually the National Board convenes to conduct corporate business and to elect officers.

The officer designated by the Air Force as Commander, Hq., CAP-USAF, is automatically the National Commander of CAP. As such, he functions as the chief executive officer or administrator of CAP.

The Chairman of the National Board and the Vice Chairman, both CAP members, are elected by the board membership. The National Finance Officer is elected by the National Executive Committee, as is the National Legal Officer.

Since the National Board usually convenes only once a year, it needs a subordinate governing body to carry through its programs and one which can convene as the need arises. The National Executive Committee (NEC) serves this purpose. The NEC is composed of the Chairman of the National Board, the National Commander, the Vice Chairman of the National Board, the National Finance Officer, the National Legal Officer, and the eight regional commanders.

The NEC might be considered the "work horse" command element since it has the responsibility for reviewing reports appropriating and raising funds, supervising the corporation's investments, establishing trusts and appointing trustees, negotiating contracts, approving budgets, accounting for expenditures, etc.

The National Commander wears "two hats." Not only does he act in the capacity of civilian corporate commander of CAP subordinate structures, but he is also the military commander of Headquarters, CAP-USAF, the Air Force level headquarters responsible for liaison support to Civil Air Patrol. The USAF region and wing liaison offices are extensions of Hq. CAP-USAF. The staff of Hq. CAP-USAF also functions under its commander as the National Headquarters of Civil

Air Patrol.

You can compare the field organization of Civil Air Patrol to that of the USAF. Each has a mission to accomplish and certain territorial areas in which to operate. However, the USAF has broken its field organization into major commands which are designed to perform specific mission functions—to greater and lesser degrees. On the other hand, the CAP field organization units all have equal responsibility for carrying out the CAP mission, but they do it within certain territorial boundaries, first by groups of states and then by individual states.

CAP regions are the first level of command in the field organization structure. Commanding each region is a Civil Air Patrol Officer, usually in the grade of colonel. Each regional commander is appointed by the Chairman of the National Board. The region commander may then appoint deputy region commander and a staff to assist him in his duties. The deputy region commander and staff may perform those administrative duties peculiar to region level, but the region commander retains command responsibility for all CAP activities within his region.

The CAP wing is the command level assigned to each state. Wing commanders are elected by the National Executive Committee and usually hold the grade of CAP colonel. Like the regional commanders, wing commanders may appoint a staff to assist them with their duties. Also helping the wing commanders are the USAF wing liaison officers. The wing liaison office serves the same purpose to the wing as does the regional liaison office to the region.

CAP wing commanders appoint the group commanders and squadron commanders within their respective wings. In large wings (usually over 100 units or where geographical boundaries dictate) sectors may be formed as an extension of the wing headquarters staff when the number of groups under the wing is too large or too scattered to permit the wing commander and his staff to exercise effective supervision.

The squadron is the community-level of Civil Air Patrol. It is the CAP's operational unit that actually carries out all of those plans and programs formulated and directed by the higher echelons. Squadrons are trained to furnish assistance to the communities, states and nation in times of national disasters, aircraft accidents, national emergencies and war. Squadrons recruit new members into Civil Air Patrol. The designation of each squadron indicates whether it is a:

Senior Squadron—composed entirely of senior members;

Cadet Squadron—composed primarily of cadets with a minimum of three senior members to meet supervisory, administrative, and training requirements in the conduct of cadet programs;

Composite Squadron—composed of both senior and cadet members and conducting both senior and cadet programs.

Flights are the smallest organizational element and are established only if a need exists. That need occurs normally in sparsely populated areas where there is an insufficient number of members to form a squadron. A flight must be composed of a minimum of eight members, three of whom must be senior members. The flight according to its remoteness, may report directly to either a squadron, a group, or its state wing—as the wing commander may direct. Each flight has as its goal the increase of membership so that it may become a squadron as soon as possible.

Civil Air Patrol units—flights and squadrons—do not come into being automatically. Each one is individually chartered as authorized by the National Board and each charter is issued only on application of a group of citizens committed to the objectives and principles of the Civil Air Patrol and on the recommendation of the appropriate wing commander. Charters are granted for one year and must be renewed annually enabling the wing commander and the National Board to insure that units reflected by the national roster are, in fact, active units capable of carrying out the CAP program in their communities.

In the final analysis, the flight/squadron at the community level is the backbone of this organization which throughout the civilized world is unique in the annals of aviation and public service.

Maj. Gen. Lucas V. "Vic" Beau, USAF, CAP National Commander during the immediate post-World War II years, greets Swiss, French and British cadets to Washington National Airport. The Civil Air Patrol International Air Cadet Exchange, which annually sees 200 CAP cadets change places with their counterparts in 20 or more friendly foreign nations, still is going strong and represents one of CAP's more coveted special activities.

Times and clothes change even in the Civil Air Patrol. Skirt length and uniform style mark these CAP female cadets visiting Lackland AFB, Tex., as circa 1950s. These girls were among the 102 honor cadets representing what then were 48 states and three territories attending the all-girl encampment at this Air Force basic training facility.

The 1950s were a period of "show and tell" for the Civil Air Patrol. Their wartime missions gone, CAP members had to demonstrate for the public that they had a peacetime role. Airlift of emergency blood supplies for the American Red Cross quickly became an important mission in many parts of the country.

A major Civil Air Patrol project in the years immediately after World War II was air marking. CAP cadets and senior members, in cooperation with the Civil Aeronautics Administration, laid out hundreds of aerial signs like this one at Albany, Ga.

You don't have to look twice to identify this as the doughty Piper Cub. During World War II these planes, known as the L-4 Grasshopper, were used in artillery spotting and battlefield reconnaissance. In CAP they became a mainstay during the 1950s for air search and cadet orientation operations.

CHAPTER 6

CAPCOM

"Without reliable communications, the only thing I can command is my desk!"

The speaker was Air Force General Thomas S. Power, then Commander-in-Chief of the Strategic Air Command, and he was stating the case for the dedicated radio communications network he felt necessary to provide command and control of SAC's far-flung intercontinental missile and bomber forces.

What to Tommy Power was a mere statement of fact has since become an often repeated, profound observation on this primary imperative to effective command and it has a special significance in the case of the Civil Air Patrol.

Perhaps the single capability which sets the CAP off from every other volunteer search and rescue and emergency service organization is its nationwide radio communications network of more than 17,000 fixed, mobile, portable and airborne stations. Where virtually every other organization—except full-time government agencies such as the Armed Forces—must rely on outside communications for command and control and in so doing further tax established emergency communications resources such as the Radio Amateur Civil Emergency Service (RACES), the Amateur Radio Public Service Corps (ARPSC) and the citizen's band REACT organization, CAP not only can handle all its own operational communications requirements, but also usually has sufficient capacity to provide backup to other disaster relief organizations such as the Red Cross, Salvation Army and the National Guard.

Nowhere has this unique CAP resource proven itself more dramatically than in the middle Atlantic states during February 1957.

Pelting rain continued to come down as it had for more than 72 hours. Three days of downpour turned the brooks into creeks, the creeks into rivers and the rivers into seas of raging destruction in western Virginia, southern West Virginia and southeastern Kentucky. Rising water forced the Guyandot, Clinch, Kentucky and Cumberland rivers out of their banks. And the rain kept up.

In a 15,000-square-mile area dominated by rugged hills, narrow valleys and scores of soft coal mines, the cry "flood" threw terror into the hearts of the mining folk. For the next five days, they lived in terror. Death and destruction reigned in the shadow of the Clinch Mountains.

The first indication the rest of the nation had that a national disaster was in the making came from a news report in the Louisville (Kentucky) Times the afternoon of Tuesday, January 29. The news report warned of rising waters and included a U.S. Weather Bureau statement to the effect that "no relief was in sight."

Two days later the Louisville Times reported:
"Pikeville — Water stood three to nine feet deep in the business district. No milk and little bread is available. Prisoners from the City Jail were released when water from Levisa Fork poured in . . .
"Paintsville — The town braced for what was expected to be a record flood stage of 45 feet today . . .
"Barbourville — At least 200 families, an estimated 700 persons, were evacuated or left their homes before they were reached by the rising Cumberland River . . .
"Hazard — Flood damage from the raging Kentucky River was expected to exceed $5,000,000 (it finally exceeded $20,000,000) according to Dewey Daniels, Hazard banker and state Republican leader. Radio reports said a hospital with some 50 patients was without electricity . . .
Corbin — The town requested 500 cots, 1,000 blankets and two field kitchen units."
At the same time, flood waters were wreaking havoc in both Virginia and West Virginia.
Almost immediately, state and Federal relief agencies swung into action. The American Red Cross began to marshal its facilities. The Federal Civil Defense Administration ordered its field office to give all possible support. Governor A. B. "Happy" Chandler of Kentucky and Governor Cecil H. Underwood of West Virginia asked President Dwight D. Eisenhower to declare the region an emergency area.
Even earlier, however, CAP began emergency operations which later were to prove vital to the overall relief effort in "Operation Jupiter" as the flood mission was to be called.
Less than a half hour after the Louisville Times hit the street that January afternoon carrying the story of the expected flood, LTC Huston H. Doyle, 43-year-old airways operations specialist who commanded the Civil Air Patrol's Kentucky Wing in his spare time, prepared a message to all units in Kentucky. Especially to London and Hazard, in the path of the flood waters, he signalled:
"Request your squadron provide all possible assistance . . . "
The rolling tide of disaster unleashed by the heavy rains moved faster than expected. As Doyle was preparing his message, MIDDLEGROUND ONE SIX, communications station of the Hazard Squadron, came on the air with a desperate plea for help.
Lt. Bill Roll, 26-year-old Army veteran who won the family bread as a gas serviceman, was at the microphone. The building at the Hazard airport which housed both the CAP headquarters and the State Police was already under water. Roll was operating from his home on high ground and on power supplied by a small gasoline generator kept for emergencies. His message, which was picked up as far away as Atlanta, Ga., said:
"HAZARD BUSINESS DISTRICT COMPLETELY WIPED OUT. FOUR FEET OF WATER IN PEOPLE'S BANK. FIVE MILLION DOLLARS DAMAGE. SEVERAL HUNDRED

PEOPLE OUT OF HOMES. WILL BE SEVERAL DAYS BEFORE CONTACT POSSIBLE BY ANY OTHER MEANS THAN CAP RADIO NET."

The message was signed by Dewey Daniels, Kentucky state Republican chairman. The Kentucky Wing immediately forwarded it to the office of Governor Chandler. Meanwhile, CAP communicators Capt. J. R. Patterson and Lt. Peggy Wade who had picked up the call in Atlanta, Ga., also were forwarding it to the Kentucky governor.

During the next four days, MIDDLEGROUND ONE SIX was the only contact with the stricken community. When Army engineer companies from Fort Knox broke through 48 hours after the first desperate message, they found Roll still on the job. With the CAP station in Hazard in command of the situation, the Army was able to move its own portable equipment on to Pikeville, another flood-paralyzed community cut off from the world.

Hard on the heels of the Army engineers came a CAP relief communications team from Louisville—Capt. Ken Galloway and S/M Wesley Jones, a minister of the Church of Christ. For the first time in two hectic days, Bill Roll grabbed some welcome shut-eye.

From Wes Jones came a graphic first person account of the devastation in the Kentucky community.

"We found the airport," he reported, "covered with six to eight inches of silt and completely unusable. More than half the 10 planes were damaged including the CAP aircraft. The airport building which housed both CAP and State Police headquarters had been under six feet of water at the peak of the flood and all equipment was damaged.

"Three thousand were homeless in Hazard, 6,000 in Ferry County. Some 4,000 were being fed by the Red Cross field kitchens. Other groups of children and adults were trying to cook food out in the open over stoves which had been recovered from demolished homes. A family subsisting on C-rations provided by the Army, shared their food with us.

All drinking water was contaminated. Many men, women and children were scavenging scrap lumber and debris from the river banks where houses swept along by the raging waters had broken up against the bridges. From these meager scraps they built lean-to shelters for their aged and young. Families who still had homes were feeding and caring for as many as four other homeless families. Their food and supplies were running out. For two days they had hoped the weather would break and let the helicopters get in with blankets, food and medicine."

Meanwhile in Louisville, Colonel Doyle was busy arranging for a high-priority airlift of serum and vaccine to the flood area. CAP lightplanes, picking up the life-preserving fluid at Lexington and Louisville, transported it to London where a combined disaster operations headquarters had been set up. Here the precious vials were put on helicopters for the last leg of the journey into the area where nature had run wild.

The first two days of the emergency, Doyle made his headquarters at the London, Ky., airport where CAP Maj. Roscoe Magee and the London squadron had been on duty since Lieutenant Roll's dramatic message went on the air. Through Lt. B. L. White, the squadron chaplain and a ham operator (W4UVH), the London CAP communicators maintained contact with Kentucky hams who also pitched in with an assist to their troubled state.

Going on the air at 1930 CST on January 29 MIDDLEGROUND EIGHT, the London Squadron headquarters station, stayed on the air continuously until 0245 CST February 2. At times, atmospherics prevented direct contact with Hazard and a relay was set up with BLUE CHIP ONE THREE, Tennessee Wing, and RED STAR FIVE, Georgia Wing.

London CAP members handled heavy traffic for the U.S. Army Aviation section set up at the airport to control the 14 huge helicopters sent in from Fort Knox, Ky., Fort Meade, Md., and Fort Belvior, Va. Billeting and office facilities in addition to communications support were provided the Army by the London Squadron.

In Hazard and in a lesser degree in other Kentucky communities, the Civil Air Patrol radio net carried the brunt of emergency traffic for all the participating agencies—the American Red Cross, the U.S. Army, Civil Defense, the State Police, the United Mine Workers Welfare and Retirement Fund which operates several hospitals in the area, county and state health, Roads and Engineering commissions, news agencies and private citizens. Activities of some 14 Army helicopters in the area were coordinated via the CAP radio net. In addition, the CAP handled its own operational and administrative traffic and carried regular weather advisories from the U.S. Weather Bureau in Louisville.

When Republican Chairman Daniels in Hazard wished to send a direct appeal to Kentucky Senators Thruston Morton and Sherman Cooper in Washington, it was relayed from MIDDLEGROUND ONE SIX in Hazard through Tennessee, West Virginia and National Capital Wing stations to the nation's capital.

The exchange of messages took less than two hours even under the severe emergency conditions and focused the attention of the Kentucky law makers on the CAP's part in Operation Jupiter. The result was a letter from Senator Morton to Maj. Gen. Walter R. Agee, then CAP national commander.

"This letter is to compliment you on the outstanding work done by the Civil Air Patrol during the recent flood disaster in Kentucky. The work of the Civil Air Patrol with emergency communications and airlift during times of such disaster has always been meritorious. The job done in this instance in maintaining contact with flood-stricken Hazard, Kentucky, as well as other phases in which your organization helped certainly was exceptional service for which the Civil Air Patrol deserves the highest credit.

"I'm very impressed with the help to my State of Kentucky and know your organization will continue the good work."

Operation Jupiter turned out to be one of the largest missions in the history of the CAP's Kentucky Wing. At its peak several hundred of CAP's civilian volunteers—each one taking time off from his job or business, mostly without pay—were manning the 18 fixed and mobile radio stations, the relief teams and the aircraft. They weren't alone, however. In Virginia and West Virginia, their counterparts were doing their share to stem the tide of death and destruction.

At almost the same time Bill Roll was telling of the plight of Hazard the first word of immediate danger in Richlands, Va., was coming in from S/M Mack Blankenship (BLUE FLIGHT SEVEN) of Bandy, Va., six miles from Richlands in the heart of the soft coal fields. Relayed by other BLUE FLIGHT stations, it was received at Hampton by Capt. Mildred Hicks, attractive wife of CAP Maj. Douglas Hicks, Virginia Wing director of Communications. Mildred, who admitted to some "30-odd" years, was an ardent CAP communicator and kept BLUE FLIGHT THREE—Virginia's alternate net control station—on the air when her husband was working at his full-time civilian job as an electronics engineer with the National Advisory Committee on Aeronautics at Langley Air Force Base.

The message which began Operation Jupiter for Virginia read:

"RICHLANDS UNDER WATER. SEVERE DAMAGE EXCEPT IN BUSINESS DISTRICT. NEED BLANKETS, FOOD AND SHELTER."

LTC Alfred Nowitsky, then deputy wing commander for Virginia, immediately ordered the entire statewide CAP organization on 24-hour alert.

In Richlands, Maj. Grady Dalton, commander of the Richlands Squadron, in private life the vice president and cashier of the Richlands National Bank, already had his unit at work aiding in the evacuation of citizens from flooded areas of the town. Two of Dalton's mobile cars were on the air maintaining contact with CAP stations outside the flooded region.

Another of the Richlands mobile cars operated by Lt. Burkley Whited was on its way in from the Whited home in nearby Swords Creek. It was some time, however, before the 48-year-old carpenter reached Richlands. Whited got as far as Reven, Va., when he found his way blocked by flood waters. Turning back, he found the flood had cut him off from behind. Some minutes later other CAP stations in the area heard Whited report:

"I'VE GOT A MILE AND A HALF OF ROAD AND NO PLACE TO GO."

He spent the night on a low ridge between two roaring torrents of water relaying messages for other CAP stations maintaining the long vigil.

One of the first orders issued by Colonel Nowitsky was to the Tazewell Squadron—the CAP unit closest to Richlands. Immediately, three mobile radio cars operated by Lt. Sam Evans (GREEN

FLIGHT ONE TWO THREE), S/M Aubrey McCracken and Lt. Carless Chaffin were dispatched to the aid of beleaguered CAP forces in Richlands. S/M Walter Blankenship (BLUE FLIGHT NINE) was assigned to coordinate the activities of the mobiles. During the periods he was to be on duty in his capacity as a Virginia State Trooper another CAP communicator, Lt. Luther Mercer, was to back Blankenship up.

Neither of the three mobiles were able to find a surface route open into the stricken community. After several tries it was decided that Chaffin and McCracken would return to their base of operations. Evans planned to continue his search for a way into Richlands.

Over the radio Evans told Blankenship he would stock his outboard motor boat with food and other supplies and would make another try. At home, Mrs. Evans overheard the conversation on their monitor receiver. By the time he arrived at the house, she had stripped the family pantry loading every available item of food into the boat along with blankets and warm clothes. Stopping only for a word of thanks, Evans hitched the boat trailer to his mobile radio car and headed back toward the swollen Clinch River.

He might have a chance, he reasoned, to get through overland if he tried the many narrow winding mountain roads in the area. Perhaps one of them would be open. Heading down Baptist Valley, he found the bridge under water. Another road, another and still another were under water when he tried them. On his last try before taking to the boat, he found an open road. Reporting to Grady Dalton in Richlands, Evans found it had taken him two hours to go 17 miles. He found also that the route he used became impassable almost immediately after he used it. It was two days before Chaffin, McCracken and the Tazewell CAP land rescue teams got through to relieve him and Dalton's weary men.

Three separate communications facilities were used when children attending the Richlands and Cedar Bluff schools were marooned by high water. The schools were warm. Food was available in the cafeterias to last through the emergency period but the problem of notifying their worried parents seemed almost impossible. School authorities had access to State Police radio but the law enforcement agency had no way of relaying the word to the hundreds of families in the area. Radio station WRIC could broadcast the good news if word could be passed to them. The State Police, however, had no means of contacting the radio station. CAP radio again came to the rescue. A CAP-operated portable station had already been set up in the studios of WRIC permitting their newsmen to keep track of rescue and relief operations. State Police cars outside the flood area picked up the message relaying it to Blankenship, now at BLUE FLIGHT SEVEN. Blankenship sent word to WRIC and the radio station put the information on the air. Relief replaced fear in many Richlands and Cedar Bluff homes.

When a rumor spread throughout the area that the Maxwell Dam had been breached, terrified citizens began to panic. Again the radio hook-up to WRIC paid off. Blankenship was asked to check. He checked with Mercer who was then off duty and was keeping his CAP station on the air. Via State Police radio, Mercer checked on the dam. His reply to Blankenship:

"MAXWELL DAM NOT BROKEN AND NOT LIKELY TO BREAK. STOP RUMORS!"

The commercial radio station put the word out on the air and general panic was averted.

Meanwhile, Sam Evans' relief after 48 hours of duty in Richlands was short lived. Sam went home and dropped exhausted into bed. The next day he planned to return to his job as a telephone inspector for the Pocahontas Fuel Co. At Bluefield, W. Va., However, a chain reaction was beginning that was to demand more hard work and personal sacrifice from Evans.

Most of that night, CAP Capt. Jim Cheek, a 32-year-old Army veteran and his wife, Lt. Norma Jean Cheek, were busy moving emergency traffic. Their powerful LOWLAND FOUR FOUR, alternate net control for the West Virginia Wing, blankets most of western Virginia also with a strong signal. It proved a perfect relay station carrying traffic also for the Kentucky and Tennessee wings. Cheek, a salesman for the Meyers Electronics Co. of Bluefield, left the next morning instructing his wife to operate the station. He began a business trip to nearby Grundy, Va., just across the state line.

At Oakwood, Cheek was turned back by State Police who said that the route was closed by flood waters. Checking the situation Cheek found that there were apparently no open routes to Grundy. Returning to Bluefield, he went on the air with a report of the situation to BLUE FLIGHT THREE.

Colonel Nowitsky, Virginia Wing mission commander, immediately checked with all state agencies in Richmond and found that there was no contact with Grundy nor had there been contact for more than 12 hours. In a matter of minutes, Sam Evans was under orders to proceed toward Grundy keeping Wing Headquarters advised of his progress. The State Director of Civil Defense asked Evans for an evaluation of the situation if and when he got into the isolated community.

Lieutenant Chaffin (GREEN FLIGHT FOUR FIVE) was sent to Short Gap., Va., a high point on the Buchanan County line, to act as a relay. McCracken and Mercer were detailed to assist Evans and another relief mission was on. It took the mobile radio cars and the accompanying ground rescue team vehicles until 10 P.M. that night to get into Grundy and then only with the aid of heavy equipment of the State Highway Department. Evans reported immediately to the office of Mayor W. B. Raines and asked for an assessment of the situation. He then sent this message—the first contact between Grundy and the outside world in two days:

"ONE ROAD NOW OPEN ONE WAY. THE ONE WE USED

TO COME IN ON. NEED CLOTHING AND BEDDING FOR 100 FAMILIES. WATER NOT CONTAMINATED. POWER BACK ON. WATER DROPPED FROM 30 FEET ABOVE NORMAL TO 10 FEET ABOVE NORMAL. TEN THOUSAND MINERS OUT OF WORK. NEED BAILEY BRIDGES TO MINES. NO LOSS OF LIFE. NO OTHER COMMUNICATIONS AVAILABLE."

When Cheek asked Evans if he could handle the situation in Grundy, the CAP citizen-turned-rescuer replied:

"JUST TELL MY WIFE I'LL BE HERE UNTIL SUNDAY AND THAT I'M ALL RIGHT."

Meanwhile, West Virginia was having its own troubles with the Guyandot River in Logan County. CAP Maj. James Singleton was the coordinator for Civil Defense for the West Virginia Wing. He also was Logan County's CD director.

"We had had experience with disaster in Logan County for a long time," he later explained. "Mine explosions, floods, complete disruptions of all types of communications caused by forest fire and snow and ice storms which cut us off from the world completely.

"Because of the continual rainfall for a 72-hour period we—the CAP—dispatched a mobile radio car to the headwaters of the Guyandot. We checked rainfall and water level in the tributaries also. This was Monday. From past experience we determined that the river would begin approaching flood stage about 10:30 Tuesday morning. The Logan Squadron immediately made plans to meet the emergency.

"Communicators were alerted and were warned to have their mobiles moved to high ground out of danger from the water so that they would be usable if and when the flood struck. Four fixed stations and 14 mobiles went on the air.

"We sent CAP mobile cars through the probable high-water area warning citizens to evacuate to high ground and assisting in evacuation wherever possible. Lt. Raymond Chapman (OVERLAND TWO SIX) alerted his area, Champanville. Man, W. Va., was alerted via radio and CAP members there began warning the population."

Now the Red Cross, Civil Defense, State Police and county law enforcement agencies took over the actual disaster assistance work while CAP stood by to provide them with emergency communication. Singleton began acting in his capacity as county CD director. He massed the joint facilities of all the relief agencies and notified Governor Cecil H. Underwood of West Virginia asking for state assistance.

At the height of the flood Major Singleton felt the necessity for a survey of the rest of Logan County. Crossing a swinging footbridge to reach the airport and his plane, a Stinson Station Wagon, he made a complete reconnaissance of the situation. On his return, the frail footbridge was wiped out less than an hour after he crossed it.

Jim Singleton did a good job in Logan County. The efforts of the Civil Air Patrol and Civil Defense, under his leadership, paid off in lives and property saved. He, however, was yet to learn of the personal sacrifice he made.

From Logan, W. Va., Singleton was but one of the many CAP volunteers who for more than 15 years had found that service to one's fellows usually means personal sacrifice. When the flood in Logan was imminent, Singleton moved one of his two personally-owned CAP mobile radio cars to a location where he thought it would be safe and available for use. The car subsequently was found under water. To make matters worse, in the press of his CAP and CD responsibilities, Singleton completely forgot his own business—commercial and neon signs. More than $2,500 worth of equipment and stock was wiped out by the raging flood waters which exceeded by six feet any previous flood in the history of Logan County.

In the aftermath of Operation Jupiter, Kentucky Governor "Happy" Chandler put it in perspective this way in a telegram to General Agee:

"Kentuckians are deeply grateful to the Civil Air Patrol for its assistance during the recent disastrous flood in eastern Kentucky. CAP members and their radio communications system performed nobly in helping protect lives and property. CAP radio at Hazard, using emergency generators when the city's power system failed, sent out first calls for help from the stricken community. Then Hazard radio working with CAP radio units in London, Middlesboro, and Louisville dispatched messages which brought in food, clothing and medicines. Throughout the flood crisis every Civil Air Patrol member involved performed magnificently and they have certainly earned our undying gratitude and esteem."

From Republican State Chairman, Dewey Daniels: "The CAP did a wonderful job here (Hazard) and without them we still would not be out of the mud. For the first 72 hours, CAP was our only line of communications."

Mayor Raines of Grundy, Va., said: "The Civil Air Patrol did a wonderful job. Shortly after the flood struck the CAP moved in to provide us with our only means of communications—and stayed on the job as long as it was needed."

The combination of a dedicated communications capability, the flexibility of general aviation aircraft, and the availability of a wide variety of special-purpose surface vehicles has, in many areas, carved CAP a special niche in statewide disaster organizations. This is especially true of those states bordering the Gulf of Mexico and the southern portion of the Atlantic seaboard where destructive hurricanes are a yearly threat and in the south central and mid-continent states of the "tornado belt."

In such areas, state governments as well as the citizenry not only expect CAP to be ready to assist, they have become accustomed to an immediate and automatic response from the blue-clad troops of the Air Force's auxiliary. For instance, the same portable gasoline-powered electric generators that make CAP's communicators independent of commercial power in a disaster area also are looked upon as a capability-in-being for emergency power for field medical facilities, hospitals and refugee shelters. In many smaller cities and towns in these states where nature regularly and frequently wreaks havoc, communities without National Guard units and where the local economy does not permit the luxury of stand-by civil defense/disaster relief forces, CAP has become not only the last resort but also the first.

In addition to the so-called standard emergency situations where CAP radio communications comes into play, there also are many unusual ones. Such was the case in the Window Rock and Genado areas of Arizona in the area covered by the Navajo Indian reservation. The reservation comprises some 24,000 square miles of the state, most of it arid, desert land. Roads and highways are few, the villages and isolated hogans are scattered throughout the desolate area. In the early 1950s, telephones were "as scarce as hen's teeth" and a real problem had come to the surface. Where serious sickness or injury brought a Navajo down it more often than not ended in the Indian's death. Modern medical care was virtually inaccessible to them, not because the facilities were non-existent in Arizona but because the Navajos had neither the communications to summon help nor the transportation to get their ill and injured to a hospital.

This situation had long bothered Arizona CAP Wing Commander Col. Dines Nelson, but now he thought he had an answer. CAP had the capability to train communicators from among the Navajos themselves. This also was during one of those periods when a large amount of communications equipment was available to CAP from the military, equipment which had become obsolete with the explosion going on in electronic technology but which, in many cases, was ideal for the CAP mission. The thing to do was to organize an all-Navajo CAP squadron on the reservation—this would make them eligible to receive the surplus equipment—and then train them in its use.

Nelson and CAP Maj. James Bickle began the effort on a modest scale, you might say as an experiment since there were many unknown quantities involved. But what began as an experiment soon mushroomed into a full-scale operation. Before long, there was not only an all-Navajo squadron added to the rolls of the Civil Air Patrol, there was an all-Navajo group with three squadrons under Bickle's command. It didn't take the Navajo tribal leaders long to recognize the validity of Nelson's idea and by 1956, the group was 130-strong, CAP communications stations dotted the reservation along with 32 airstrips some of them built by the Bureau of Indian Affairs, the first built by the Navajos.

The Navajos were fascinated by the "wind that speaks" and it wasn't long before medical advice and attention or, where required, evacuation to a hospital were just a CAP radio link away. During the first nine months that first squadron was in existence, 14 lives were saved.

Today, Civil Air Patrol's nationwide communications capability includes some 17,000 stations operated by members together with a 61-

station net linking the Headquarters CAP/USAF with the eight regional and 52-wing Air Force/CAP liaison offices. At the liaison office level these stations operated by regular Air Force or Department of Defense civilian personnel assigned have the capability and are authorized to operate on the frequencies assigned to CAP as well as those assigned exclusively for Air Force use. This permits the passage of traffic from the National Commander directly to the appropriate regional or wing commanders and their staffs. Daily net operations are carried on between Headquarters CAP/USAF and the regional liaison offices on seven and 14 megahertz frequencies using single sideband equipment and voice procedures. Regional liaison frequencies assigned range from four to 20 megahertz depending upon areas to be covered and the geographical location.

Civil Air Patrol unit headquarters and alternate stations, mobile radio vehicles, aircraft and portable stations operate on additional Air Force frequencies allocated to CAP for that purpose. They include frequencies in the two megahertz, four megahertz and 26 megahertz high frequency (HF) bands and a pair of very high frequency (VHF) channels. Operation on the lower bands employs single sideband equipment. The 26 megahertz channel used amplitude modulated (AM) equipment limited to the same power as the neighboring Citizens Band and representing identical communications limitations. Since it first became available in any quantity, VHF equipment used by CAP has been of the AM variety but with the development of frequency modulated (FM) equipment of relatively low cost and high reliability a gradual transition to FM was ordered. That transition is expected to be complete by 1975.

Frequencies allocated both to the USAF/CAP liaison nets and to the Civil Air Patrol nets are among those reserved for the U.S. military forces as part of the inter-governmental agreements arrived at within the structures of the International Telecommunications Union (ITU). CAP's access to these frequencies comes as the result of its designation as the civilian auxiliary of the Air Force and through an agreement between the Department of Defense and the Federal Communications Commission (FCC). While CAP's frequency allocation and its operational procedures are derived from the military, its stations and operators are governed by FCC regulations and as such carry FCC licenses. CAP communicators therefore must not only meet the requirements of their parent service, but also those of the FCC which often are even more stringent. This dual requirement has resulted in a degree of communications discipline and an adherence to establish procedures second to none in the communications world.

In addition to the Air Force frequency allocation for CAP use, two channels in the Aeronautical Band—122.9 and 123.1 megahertz—have been authorized for CAP use in search and rescue operations. These frequencies are common to all U.S. aircraft and the equipment they carry. This permits both corporate aircraft and member-owned aircraft to conduct air-to-air and air-to-ground communications with properly-authorized fixed and mobile SAR stations using the aeronautical communications equipment normally on board. Equipment for such fixed and mobile SAR stations must, however, be "type certificated" and meet the specifications of the airborne equipment authorized for aviation use. Equipment for CAP's normal point-to-point and fixed-to-mobile communications meets high technical standards but does not have to meet these aeronautical specifications and thus the cost is considerably reduced from that of the type-certificated gear.

The Civil Air Patrol communicator represents a wide range of background, technical capability and experience. They all have certain things in common, however. They must hold an FCC Restricted Radiotelephone Operator's Permit, they must have studied and passed with a grade of 100 percent a comprehensive 100-question examination on the principles and doctrine of CAP communications and they must strictly adhere to established on-the-air procedures based on those developed by the Air Force itself.

In communications or electronic experience and capability a cross section of CAP radio operators ranges all the way from the member who had never communicated by radio in his life before completing the prescribed course of study and being awarded his CAP Radio Operator's Permit (ROP Card) to individuals holding FCC First Class Radiotelephone Permits and employed as commercial broadcast engineers. Many FCC-licensed amateur radio operators as well as technicians employed in various phases of electronics also are CAP communicators. These highly proficient types provide the professional guidance and the technical know-how that has made and keeps CAP's communications capability what it is. But, more often than not when the chips are down and in typical military jargon "the balloon goes up," it is the housewife communicator, the retired "old-timer" or the shut-in who sees the first action. They are the people whose personal commitment permits them to man their radios while CAP breadwinners are earning a living. They are the ones that make CAP's national radio communications capability a 24-hour-a-day, 365-days-out-of-the-year reality.

Communications officer checks out a ground team member and his radio-equipped jeep during a training mission. Regular field exercises in search and rescue and disaster relief keeps CAP personnel on their toes and ready for the real thing.

CHAPTER 7

Samaritans With Wings

About 60 miles south of Brownsville, Texas, along the Gulf Coast of Mexico, you find a long peninsula running parallel to the beach. The barren sand is about a half mile wide. From time to time the natural breakwater is broken by a pass connecting with the open sea. As you travel farther south the passes become wider and more frequent and the strips of sand shorter, creating a series of little islands.

The waters in the passes and around these sandy islands abound in reds and salt water speckled trout. It's fisherman's paradise. Cast into the surf or bait-fish the holes just off the water's edge and you can pull in a hundred pounds a day. It's Tom Robertson's country and this is his story!

An old friend, Herman Watkins, and I began fishing the passes more than 20 years ago. When we loaded my new Ford pick-up that sunny June morning in 1954, piled my 10-year-old son, Bill, in between us and headed southeast from Mission toward Brownsville and the border, it was like scores of similar fishing trips before.

We got started about 7 a.m. About two hours later we were in Brownsville, where we planned to get gas, groceries and clear Mexican customs. As usual I put in a call to the U.S. Weather Bureau to check on conditions in the Gulf. It's a tricky body of water and a lot of violent tropical disturbances find their birth there. Today, however, clear weather with no unusual weather activity was the word.

Herman packed a 300-pound block of ice into the chest. Bill and I loaded a week's supplies into the truck and we cleared customs. I checked the mileage on the speedometer. Just 8,000-miles. The Ford was hardly broken in. This trip would do it good.

From Brownsville we drove down the highway to Washington Beach. Here we headed down the hard sand at the water's edge. It was 2 p.m. when we finally hit the peninsula. It's another 25 miles to the first pass so we decided to take our time for the rest of the afternoon, stopping at likely looking spots to surf cast and fish in the holes.

Night was falling when we crossed the first pass. We drove the last few miles to the second pass and chose a camp site in the gathering dusk.

The weather usually is beautiful in this part of the country. Making camp demands only putting up a tarp for shade during the day, unfolding the camp cots and hitting the sack.

The next day, Wednesday, we caught our hundred pounds and more. Some of the catch supplemented the larder that night. The fresh fish sure hit the spot. Bill consumed so much I could swear he was going to pop.

"This is the life, Tom," Herman remarked as he leaned back on his elbow in the still warm sand, puffing an after dinner cigarette.

I was just about to agree when I got the shock of my life. One minute we were sitting comfortably on the soft sand, lazily recollecting a satisfying day of angling. The next minute we found ourselves in two to three inches of water. Bill jumped up hurriedly, looking ruefully at the seat of his chopped-off blue jeans.

"Where did the water come from, Dad?" he asked.

"Darned if I know," I told him, turning to Herman. "What do you make of it?"

"It's a new one on me," he replied. "Get a flashlight and let's take a look."

I rummaged around for a minute in our gear and came up with the big three-cell light I always carry. Holding Bill's hand, we began walking slowly down the beach, shining the beam of light ahead of us. As far as we could see the waves were washing up on the beach over the tracks the pick-up had made on the trip down.

"What's going on?" I asked.

"It's beyond me," Herman answered, a worried note coming into his voice. "I've never seen the tide run this high. One thing's a cinch—the truck is going to be hopelessly bogged down if we don't get it up on something. That sand gets treacherous when waves begin breaking over it."

We turned back swiftly to the camp. The water had receded. The sand around the camp site was beginning to drain dry again, but the truck wheels had already sunk several inches. While Herman and I gathered some drift wood and, using Coke cartons, laid a solid surface on which to run the truck, Bill policed up the camp site, putting all the loose articles in the truck.

We built our makeshift platform on the highest spot of ground in sight and drove the truck up on the support.

"I don't think we'll get washed away during the night," Herman observed as we returned to camp. "The tide is just running unusually high."

Agreeing, I said to Bill, "We'd better turn in, son."

In a few minutes, Herman was snoring and Bill's breathing was slow and quiet—the sleep of youth. I soon dropped off, but something in the back of my mind wouldn't let me find a deep, restful sleep. Twice that night I awakened with a start to find the water again lapping at the cot legs.

Daylight was just seeping in from seaward when I awoke. The skies were overcast and a light drizzle was pattering on the tarp over our heads. The wind was calm, but the tides were running even higher than the night before.

Herman and I took a look down the beach and agreed that it would be impossible to get the truck out with the water as high as it was. We ate a light breakfast and a cold one, then sat around till the water began to go out about 10 a.m.

"Let's try and get the truck out," Herman said. "We can head back up to the first pass where the peninsula is wider and there is higher ground. No use having a wet camp all the time we're down here."

His reasoning smacked of common sense so I agreed. We were both wrong, however, as we found out once the wheels of the truck touched the water-softened sand. In seconds the rear wheels dug themselves in up to the axles. One thing for sure, we weren't going anywhere in the pick-up this day. The rest of the morning was a story of back-breaking labor and disappointment as we vainly tried to get the Ford back up on the Coke cartons and driftwood.

Up and down the beach we scrounged for more driftwood to put under the wheels—anything to give them traction. All to no avail.

At Herman's suggestion we tried to jack the truck up and put wooden supports under it. No dice. Just when it seemed we were getting somewhere the jack broke. From that point on the pick-up was of no use to us.

It must have been during the time we were working with the truck that the rain began to fall steadily and the wind began picking up. We didn't notice it at the time, but when the jack broke and we abandoned the idea of getting the truck out we began to sense the change.

"It's going to be pretty nasty," I said as we got under the tarp and began to nibble at a cold lunch.

"Shouldn't get too bad," Herman replied. "We can still fish while we wait for the squall to blow over."

For the next two or three hours we fished when the rain would let up and dashed for the tarp each time it returned. It must have been 3 p.m. when the rain began falling by the bucketful and the winds increased.

Gathering every loose article in sight, we made a run for the pickup, climbing into the cab for protection. We weren't too wet. Fortunately, you don't wear many articles of clothing fishing in the tropics. When you get wet, the less you are wearing the sooner you dry out. Bill had on his cut-off jeans and a light jacket over his undershirt. Herman and I were wearing swim trunks, shoes and our hats. I managed to salvage a light jacket also.

We had been in the pickup for a half hour or so when the tarp tore away and Herman remarked, "It's a good thing we didn't stay out there."

The truck was parked heading south, its rear toward the northeast winds now approaching gale force. We didn't know it at the time, but this was the first caress of Hurricane Alice, which swept up the Gulf Coast inflicting millions of dollars in damage.

The swells driven by the screaming wind now were striking the truck from the rear. Each time a three to four foot wall of water would strike the truck, more gear would wash away. Slowly but surely, the water cans, bed rolls, cots and food boxes disappeared from the truck bed. That Ford sure was solid; no water found its way into the cab. It was only a matter of time, however, until the storm-tossed waters would catch up to us. Our real trouble began when a particularly vicious wave drove the ice box lid against the rear window with the power of a piledriver. The clear glass cobwebbed from the blow but held.

As if the blow on the window was a signal, the winds calmed. The lull that hit us was almost as psychologically disturbing as the incessant battering of the waves. The water was now knee-deep around the truck, even in the backwash of the waves.

The winds began to shift to the southwest. As the first waves washed over the radiator, I grabbed Herman, yelling:

"We can't stay here. We'll drown for sure. Our only chance is to find higher ground—a dune or something."

He grimly nodded his assent. With each of us gripping Bill's hand, we piled out. We knew now it wasn't just a squall, but a full-fledged hurricane. Thank God we still had a flashlight that worked. By now it was pitch dark. In the roar of the winds, the sound of the surf was undetectable. Our only reference was blind instinct and the knowledge of what general direction would take us to higher ground.

Gradually, foot by foot, we made our way. Bill was scared. What 10-year-old wouldn't be? Actually, I guess it was lucky he wasn't older. He didn't quite realize we had a thousand to one chance of coming out alive.

I held Bill close to me for a moment while I said a silent prayer. Then, trying to keep the fear out of my voice, I spoke reassuringly to him. Somehow I kept my voice firm. Bill quickly became calmer.

By this time each gust of wind would blow us flat on the sand. The swells washed up around our waists when we could stand. Our sense of direction was gone. I knew if we kept going we might be walking into deeper water rather than toward safety.

I pulled Herman close and screamed to make myself heard over the roar of the wind:

"We can't go any farther. Sit down. I'll hold Bill between my knees. Dig in your heels against the undertow and pray. The swells will break over our heads, but most of the time we will be able to breathe."

I don't know how long we stayed there. Each wave crashed over our heads. As it receded the undertow sucked at us like a million grasping hands. Between waves we gasped for breath and Herman and I would check to see if Bill was all right. During the next few minutes I must have said every prayer I knew and a few hundred I made up as I went along. The boy took it like a trooper. He would get uneasy or be taken with a coughing spell from a mouth full of salt water, but Herman and I would reassure him and he would quiet down.

It was like a sight of heaven itself when streaks of dawn seeped in, lighting the grey, storm-tossed horizon. As soon as we could see our way to higher ground, we started out. The water hadn't receded. The winds continued at hurricane force. Now we were faced with another foe—windwhipped sand that ripped at our faces, blinding us and blistering our hands, arms and legs.

It was full dawn when we got to the one and only sand dune in sight. It was about 20 feet in diameter and five or six feet high. We dug in on the leeward side of the dune, our feet at the water's edge. Herman and I first covered Bill with sand, then ourselves to keep the chill off. We settled down to wait.

It was noon, Thursday before the wind began to go down and the water receded. As soon as it was safe, we made our way back to the pickup to see what we could find in the way of food and water.

The hood of my brand new truck was just visible above the sand. You couldn't open the doors. In the back we found the sum total of two packs of gum and four bottles of Coke. We each took a bottle of Coke and one stick of gum and headed down the beach. I hoped that a Mexican fish camp near the third pass had withstood the storm and would offer shelter and food. The fish camp was still there, just barely, and the Mexicans offered us part of the meager supply of coffee, tortillas and beans they had managed to salvage. That night their dog caught a jackrabbit. We dined in style.

Our food stopped right there, however. Instead of three of us there were now six. Bill was beginning to suffer from exposure. He would need good food and medical attention before long. At the least, it would take us three days to walk out. Bill might be dead by then for all we knew.

Twice we tried to flag down a U.S. Coast Guard plane which passed overhead. He didn't see us. Again when a private plane flew over the beach we waved like mad. The pilot only waved back and continued on his course.

As I looked up the beach, my heart suddenly gave a start. Almost on the horizon, circling over the spot where my pickup was buried in the sand, was a speck. Straining my ears, I could barely hear the drone of an airplane.

"Herman, Bill!" I yelled at the top of my lungs, "there's a plane over the truck. Wave at it, wave as hard as you can. He may come closer. He's got to!"

Almost as one, we began running pell-mell toward the plane, waving and screaming like a bunch of Indians.

As we watched, the plane stopped circling and began to gain altitude.

"Please, God," I prayed, "send him this way!"

As if in answer to my prayer, the speck grew larger and soon I could make out a bright red monoplane. It came closer and then dove over our heads. The pilot waved and we waved back. This one wouldn't go away, we knew. He had come looking for us. I thought at the time I recognized the plane as one flown by the Civil Air Patrol from McAllen, a small town near home.

Herman, Bill and I stopped, panting in the sand, and stood with up-turned faces. I looked around, but couldn't see any place the plane possibly could land. There was one tiny spot clear of water or soft sand. It couldn't have been more than 400 feet in length and only 15 to 20 feet wide.

"He can't possibly land," I told Herman, "but he will send help."

I was wrong, however. He did land. Flying that plane like only a master pilot can, he brought it in with inches to spare.

Sure enough, it was the Civil Air Patrol—Harvey Ferguson of the McAllen Squadron. We talked briefly. Harvey felt he could get off the ground with Bill and me in the rear seat. He would come back for Herman. Again the man's ability as a pilot paid off. We were airborne just as the plane's wheels seemed to drag in the water. In an hour and 20 minutes, Bill and I were safe at Brownsville International Airport. Bill was given hot food and a doctor checked him over. Harvey made a second flight back down the coast to bring in Herman.

At Brownsville I found that my wife had notified the CAP when the hurricane struck the Texas coast several hours after it struck us. As soon as it was physically possible to fly, Harvey was to have made the first search flight. The others were to follow.

Today, thanks to the grace of God and the courage of my Texas neighbors wearing the uniform of the volunteers of the Civil Air Patrol, Herman, Bill, and I can go fishing again.

To Tom Robertson, Herman Watkins and young Bill, Harvey Ferguson is a bona fide hero. Ferguson, like most CAP members, doesn't see it quite that way. To him the Robertson find and rescue merely represents business as usual for a working Civil Air Patrol member. In the records of the CAP—now in its fourth decade—the rescued and their rescuers number in the thousands.

In Rochester, N.Y., sunburned, hungry and voicing his praise of CAP, Dr. Kenneth W. Moore stepped ashore from a cabin cruiser that picked him up after he had spent nine hours adrift in an open boat lashed by 17-foot waves on Lake Ontario. The 58-year-old dentist was none the worse for his long hours in the 12-foot plywood boat but he might have been if it had not been for CAP pilot Earl Siegfried and his observer, Joseph Keable Jr. The pair manned one of the five planes launched despite high winds to search for the missing doctor.

Moore was on a mercy mission himself when he became stranded in the skiff. He was enroute to bring in Wyatt Raleigh whose outboard had been swamped a mile and a half off shore. Moore's boat also was swamped as he tried to lash it to Raleigh's craft. Raleigh ended up swimming to shore while the doctor's boat was blown out into the angry lake awash to the gunnals.

When hundreds of volunteer ground searchers failed to find 79-year-old John Tomb of Allegan, Mich., after two days of searching the rough countryside, CAP was alerted and an aerial search located the man within four hours after the first aircraft took off. The man was spotted by Capt. Michael Uramkin and observer Jerry Winter who were conducting their effort at tree-top level.

CAP's motto is "Semper Vigilans"—always vigilant—and it isn't at all unusual for an alert Civil Air Patrol unit to be Johnny-on-the-spot even before they are needed, at least insofar as the "victims" are concerned.

CAP pilot Les Bronson and his observer, Rev. Paul Carlson, found three hunters in the Alaskan wilderness even before the trio knew they were in trouble. A break-up of ice on the rivers in the Nome area caused local agencies to fear for the lives of the three men, especially when they became overdue on their return. Bronson and the reverend were the first out and soon saw the three men. Risking a landing on the icy terrain to direct the hunters to safety, the "rescuers" found that the men had just decided to prolong their hunting trip.

Shocked at seeing a plane settle into the icy wilds, the hunters hastened to the ship.

"What's the matter, buddy," one of them asked, "Lost?"

It was a different situation for CAP Maj. C. H. Widner of Yuma, Ariz., and Air Force Col. Elmer McTaggert. The target of their search, an elderly Los Angeles, Calif., man, was really lost in the blistering desert beyond Yuma near the Mexican border. When the victim was found he already had walked nearly 50 miles and was in a state of exhaustion. After dropping food and water, the fliers returned to Yuma, picked up a jeep and completed the desert rescue.

Sometimes the aerial capability of CAP isn't needed. There are situations where special-purpose surface vehicles—CAP has hundreds usually uniquely suited to the terrain where they are used, snowmobiles, weasels, dune buggies, swamp buggies—are needed. In the rugged Sierras of western Nevada jeeps are a primary means of land rescue. Adventurous David Smith, 12, son of Mr. and Mrs. Harold Smith of "Harold's Club" fame, and a friend, Hart Tisdale, were rescued after spending more than 20 hours in sub-zero cold. The boys had left home on their saddle horses intending to cross Mount Rose to Lake Tahoe. They had not, however, informed their parents of the plan. When they failed to be home at nightfall, the parents notified Washoe County Sheriff C. L. Young who promptly called the nearest unit of the Nevada CAP Wing's Washoe Jeep Squadron.

Within a half hour after the call was received the entire squadron had been alerted and the groups of jeeps equipped for mountain rescue were scouring the trails and roads into the Sierras which stretch up to 8,500 feet west of Reno. Platoons of jeeps systematically worked trail after trail in the freezing cold maintaining contact by radio with other units and with mission headquarters. Meanwhile, the air alert had been sounded and Maj. Al Butler, CAP, Reno Squadron commander, readied his aircraft and crews.

Just before dawn broke, Les Owens, one of the platoon commanders smelled a whiff of smoke and led his jeeps along the rough trail that borders Hunter Lake. In a few minutes the searchers came upon the boys huddled over a fading fire. Other than being cold, the youngsters were in good condition. The search for the two boys represented the third such problem faced by the squadron in a two-week period. Earlier, it had participated in a search for a missing plane and the hunt for two lost hunters.

Jeep units first were organized in the Nevada Wing in the early 1950s and well into the 1970s they still are going strong and in heavy demand by local authorities taking part in many search activities that never get to the aerial phase.

On the other hand, CAP's airlift capability—especially the flexibility of lightplanes in the hands of experienced aircrews—are in constant demand for a type of mission generally known as the "mercy mission."

Three hours of back-breaking work was necessary to start an Alaskan Wing L-5 ski-plane when word came that 84-year-old cannery employee Tom West was in critical condition and required immediate hospitalization. Capt. Bob Mason volunteered to make the flight in the 45-degree-below-zero weather to the 1,100-foot landing strip at Seldovia where drifts were so huge that a tractor and sled were needed to transfer the patient to the ambulance-equipped plane for the flight to Anchorage's Providence Hospital.

FAA communications personnel at Acomita, N.M., and the parents of a Hedgwick, Kan., child had reason to remember CAP as the result of two similar missions. When Carl McClain, communications station chief, became seriously ill while on duty, his assistant drove him to nearby Grants only to find the town's only physician out of town on another emergency call. The assistant then contacted the airways communications station at Albuquerque which in turn called the Civil Air Patrol. In an hour, Maj. Fred Adams was on the ground at Kirtland AFB with the patient aboard his L-5 ambulance plane. McClain recovered. At Hedgwick, four-year-old Michael Slarrow was stricken by a brain tumor. An operation at the Mayo Clinic, Rochester, Minn., was needed to save his life. LTC James J. O'Conner, deputy Kansas Wing commander, and Maj. Kenneth Chatfield, Kansas Wing USAF/CAP liaison officer, cooperated to get the child from Hedgwick to the clinic.

A need for emergency blood of an extremely rare type was met in Fredricksburg, Texas, by LTC Joseph Braun, local CAP group commander. Two hours after the call for blood had been put into the regional blood bank at Waco, Colonel Braun landed at the Gillespie County Airport with three pints of the AB-RH negative type blood. The patient, a young mother, recovered from her illness but, according to doctors, her life would have been "seriously threatened" without benefit of the blood expeditiously airlifted by CAP.

In another blood lift, Capt. Jess Allen of the Delta Squadron, Antioch, Calif., was credited with saving the life of Allan Collin of Los Angeles. Collin was stricken suddenly while in Antioch on business. The supply of blood ran out just when it was needed and the attending physician appealed to CAP. Captain Allen flew to Stockton for the blood and after 10 pints had been added to the two quarts already administered in the Antioch Hospital, his physician reported the man was "out of danger."

The life of a Thermopolis, Wyo., child was saved when Col. Thomas Knight, CAP, flew the infant, in a critical condition, to Denver, Colo. Clyde Nasaftie, a three-year-old Hopi Indian boy is alive today because he was rushed from Polacca, Ariz., to an Albuquerque, N.M. hospital for emergency surgery to remove two foreign objects from his stomach.

Four-year-old Gabe Farkas of Amityville, N.Y. was given a better chance to live thanks to the lightplane and the Civil Air Patrol. Little Gabe was suffering from megalocephalic encephalitis and had been given five months to live. Local physicians said that treatment available in Boston, Mass., might prolong his life but several obstacles to the move were present. They said that the trip was out of the question by either train or ambulance and that he might also be adversely affected by a plane trip in a commercial airliner due to the altitude the big plane would operate at and the fact it could not immediately land if the child's condition took a turn for the worse.

When the situation was brought to the attention of Col. Joseph Crowley, CAP, commanding officer of the New York Wing, he ordered a CAP L-5 liaison plane prepared for the trip. Ed Lyons, executive office of the Nassau Group, and Dan Brigham, air rescue officer on the New York Wing staff, took off with orders to land immediately if the flight should begin to affect the child. The mission proved entirely successful, however, and the child was delivered to hospital authorities at Logan Airport where airport officials waived the usual landing fee for the mercy flight.

Back in Alaska, Bob Mason was at it again. Flying his L-5 into the mountainous Big Lake region, he made an "impossible" landing and take-off airlifting Mrs. Oscar Anderson to the Anchorage hospital. Still another Alaska Wing pilot, Earl D. Smith of the Anchorage Squadron was cited for "outstanding service" in ferrying rescue workers to the scene of a spring avalanche which claimed the life of an 18-year-old youth and trapped three other skiers under a blanket of snow. The snowslide occurred on the slopes of 2,750-foot Lookout Mountain. When the avalanche was reported, Smith flew to the top of the mountain making six trips that day with rescue equipment and men, supplies, food and a toboggan for the body of Donald Koehler who suffocated before rescuers could extricate him.

Because of the inaccessibility except by air of much of its area during the winter months, Alaska has a greater requirement for missions of this kind by the Civil Air Patrol than most other areas of the United States. In fact, by agreement between the CAP and the Air Force rescue forces in our most northern state, the auxiliary's volunteers get all those missions not requiring the heavy, high speed ARRS aircraft. This operation concluded by CAP veteran bush pilot Mason LaZelle is typical:

Flying his ski-equipped L-5 liaison plane 210 miles into the barren Alaskan wilds against adverse weather conditions, LaZelle airlifted an Air Force doctor to the isolated village of Lime when the natives were stricken with influenza. LaZelle volunteered for the dangerous mission after the 10th Air Rescue Group requested assistance from CAP.

Word of the epidemic which had stricken 16 of the 40 villagers first reached the outside world when two of the residents, Katherine and Alexander Bobby, broke trail 35 miles through the snow to Stony

Village requesting Postmaster Reg White to send a message for medical aid. The two heroic residents were suffering from the disease themselves when they made the trip.

White flashed the message to Anchorage where CAP was asked to take over. The physician, Capt. Carl M. Russell of the 39th Medical Group, and the necessary medical supplies were ferried to Anchorage's Merrill Field from Elmendorf AFB by helicopter. In the face of the high winds in Merrill Pass, LaZelle and Russell took off in their light plane for Lime. Bad communications conditions prevented maintaining radio contact with the L-5 and Rescue dispatched an SC-47 to Lime to check on the light plane's safe arrival there. The SC-47 crew reported seeing the L-5 safely on the ground.

Arriving at Lime, the doctor found that many of the villagers were at Whitefish, another 35 miles away, on a trapping expedition. After treating the sick at Lime, he and LaZelle continued on to Whitefish where they found almost all of the trapping party stricken with the disease and suffering from starvation. Bears had raided the food caches at the beaver trappers' camp and the trappers had become too ill to hunt or trap more food.

Additional food and medical supplies were flown in and dropped from the SC-47. Following Russell's return to Anchorage, an ARS L-20 Beaver was flown in and three of the more seriously ill were evacuated to Anchorage hospitals. The Beaver was dispatched after LaZelle reported he thought the heavier plane could land and take off safely.

To make the hazardous trip, LaZelle had to carry five-gallon cans of gasoline lashed to the deck in the rear of the plane, as no gas supplies were to be found in the interior.

CAP was asked to take the mission because it was inadvisable to risk the heavier Air Rescue planes until the exact condition of the landing surfaces could be checked.

While individual humanitarian missions remain vivid memories for those whose lives were spared as well as for those who brought help when it was needed, such operations usually end up as items in musty record books and rarely become widely known. Another type of CAP mission—disaster relief—usually is given broad coverage by the media and in fact has brought the Civil Air Patrol much of its well-earned fame.

CHAPTER 8

Blueprint For The Future

The Nevada desert is cold in the early morning hours even in May. The 2000-odd men and women milled around to keep warm. Many were muffled to the ears while other braver souls wore only light jackets. All, however, made frequent trips to the coffee pot.

Some of them stood close to the roped off area where the television crews worked. Newsmen pounded out a last story on portable typewriters. Photographers made last-minute adjustments to their cameras. The clerks, mechanics, salesmen, bankers, housewives, and government officials gathered in little groups talking.

All of them looked out from time to time across the dry lake bed toward a tiny electric light which seemed to glow with a dreadful melevolence. From its vantage point seven miles away the light seemed to reach out, bringing to each a mixture of emotions—excitement, wonder, fear and concern.

The babel of their voices suddenly was stilled by the loudspeaker.

"The atmospheric conditions are such that the shock wave will be unusually strong this morning. All drivers are asked to lower their windows to prevent breakage and flying glass."

The abrupt announcement stirred the spirit of high adventure in some. It deepened the feeling of concern in others. There was a small flurry of excitement as drivers hurried to their trucks, busses and automobiles to obey the order. Watch hands crept toward 0500 hours. The air of tenseness increased, punctuated at regular intervals by the loudspeaker.

"It is now H hour minus fifteen minutes!

"It is now H hour minus ten minutes!

"It is now H hour minus five minutes!

"It is now H hour minus one minute!

"It is now H hour minus fifteen seconds . . . fourteen . . . thirteen . . . twelve . . . eleven . . . ten . . . nine . . . eight . . . seven . . . six . . . five . . . four . . . three . . . two . . . one . . . "

Millionths of a second seemed to drag out into hours as they waited. Some looked toward the tiny light 500 feet above the valley floor through heavy goggles so dense that the sun would seem to be the flicker from the taillight of a firefly. Others huddled with their backs to the valley, hands clasped tightly over ungoggled eyes.

It came silently. Even the rush of the desert wind was stilled. As the horrible magnificence of fissioning atoms drenched the landscape with the light of 20 suns, the 2000 voices joined in a mighty exclamation.

Each counted off three seconds—"one one-thousandth, two one-thousandths, three one-thousandth". Those with high density glasses pushed them up. Those who faced away from the blinding light turned fearfully toward its source.

Seconds crawled by. The nuclear nebulae changed to a bright orange fireball swelling and soaring upward. It forged skyward—orange to blood red, then to violet, followed by deep purple, turning finally to a dirty, greyish black. As the cloud reached the edges of the stratosphere the familiar mushroom cap of ice crystals formed, refracting the first rays of the rising sun.

Twenty-six seconds after that first fraction of an instant when the deadly blast of shattered atoms wiped out the little electric globe marking the 500-foot tower that had harbored the atomic device, the observation point trembled with a single heart-stopping thunderclap like the crack of doom. Watches stopped. Caps were knocked off. Observers exhaled audibly, then gulped a half dozen lungfuls of nerve-steadying fresh air.

Operation Cue, the 1955 atomic open shot, was over. The atom had unleashed its deadly cargo of radiation, searing heat and crashing destruction on the Nevada desert, leveling test structures, twisting radio towers, and searing the earth.

To the lightplane owner and private pilot the 1955 atomic test still is not over, however. This nuclear explosion only served to point the way to a new and dramatic role for small aircraft in civil defense. It launched an era of testing and training; and an era of developing and proving techniques whereby private aircraft and their civilian crews can bring help to the victims of atomic destruction. It placed a heavy responsibility on the shoulders of private aviation, a responsibility too great to be shrugged off.

Almost before the thunderclap had died away, a small group of civilian pilots—all members of the Nevada Wing of the Civil Air Patrol—was rushing down the gravel road to Yucca airstrip six miles from ground zero. In a matter of minutes a Stinson 165 was airborne. Aboard were two Civil Defense radiological monitors and a battery of electronic devices for measuring the deadly nuclear radiation.

The Stinson headed for an area predetermined to be one most probably affected by radioactive fallout from the explosion. For an hour it flew special patterns over the ground at various altitudes. It was tedious work for the pilot. He had to maintain a perfectly straight course on each leg while he maintained a constant airspeed and a constant altitude above the ground.

In the back the monitors were busy taking readings, carefully plotting them on special charts. They compared notes and took more readings. Just where the "hot" areas were found and how "hot" they were were classified by the Federal Civil Defense Administration and the Atomic Energy Commission. The fact that the lightplane in the hands of a civilian pilot proved a suitable vehicle for performing radiation monitoring is no secret, however. Neither is the fact that many communities jumped the gun on the atomic tests and already had begun experimenting with private planes—all operated by Civil Air Patrol personnel—in detecting radiation and plotting the hot areas.

In Chicago an airborne warning system to check radioactive fallout was unveiled earlier that year. The announcement was made jointly by Civil Defense officials and the Illinois Wing of the Civil Air Patrol. Some 20 planes owned by members of the wing already were equipped with portable radiation detection apparatus. Training in the use of the equipment and the methods of radiation detection was under way now in the seven participating CAP Squadrons.

A similar joint undertaking was announced in Oklahoma and at Oak Ridge, Tennessee. It was revealed that CAP planes had been used to measure radiation over the Atomic Energy Commission installation there as a regular method of plant safety.

In the context of the mid-1950s international situation, Maj. Gen. Lucas V. Beau, then CAP National Commander, said:

"There are two major problems with which we must deal if any American city is struck by an atomic bomb. If we are attacked we must reckon first with the problem of evacuation. Secondly, we must be prepared to bring help to those who are caught within the damaged area. In either case, private pilots with their single engine lightplanes based at hundreds of small airports across the nation will be called upon to perform a variety of missions for which their planes and training are especially adapted.

"In the case of evacuation, traffic congestion certainly will throw up roadblocks in the way of the hurrying populace. This is where the small, radio-equipped private plane with its ability to fly low and slow will be pressed into service providing aerial eyes for police agencies charged with the responsibility of keeping the flow of human traffic moving."

"Lightplanes," continued General Beau, "with their ability to operate into and out of small, improvised airfields can be used to great advantage in providing airlift for evacuating children, old people and invalids from homes and hospitals.

"After an enemy nuclear bomb has devastated an American city the lightplane and its volunteer pilot have an even more important role—that of bringing in doctors and nurses, blood, medical supplies, and uncontaminated food and water. When surface transportation is hindered by wrecked bridges, toppled buildings and abandoned vehicles, the lightplane can land in vacant lots, athletic fields, stadiums, golf courses, parks and cleared roadways. Critical supplies and rescue personnel can be brought quickly to the center of the disaster area."

A year earlier nation-wide Civil Defense test—Operation Alert—CAP planes flying into a 900-foot football field in downtown Washington, D.C., airlifted 1,700 pints of "whole blood" to within 100 yards of the Civil Defense command post. In still another training mission, CAP pilots airlifted an entire field hospital complete with 16 beds, two doctors, four nurses, first aid attendants, its own electric power plant, portable operating table and other equipment into Philadelphia after an "atomic bomb" exploded in the Navy Yard. Contemporary lightplanes ranging from Cubs to Navions were used.

In addition to the dramatic new radiation survey role, CAP planes and crews demonstrated other capabilities during Operation Cue. During a three-day period more than 70 scheduled missions were flown by CAP Cessnas, Navions, L-5s, L-16s, Howards, and Bonanzas. These included an airlift between Yucca airstrip, within the Nevada Test Site of the Atomic Energy Commission, and Las Vegas some 80 miles away. Ninety percent of all the newsreel and television film subsequently screened and most of the still pictures in daily newspapers were flown out in CAP planes. CAP also operated flights to Los Angeles to help get the story of the atomic open shot to the American public.

If this had been an actual atomic attack, CAP officials pointed out, these flights could have been carrying critical medical supplies.

All aerial photographic missions performed by the Civil Defense Photo Group were flown in CAP planes or in Bell and Hiller helicopters loaned to CAP for the tests by the manufacturers. These missions were controlled by means of CAP's own radio network set up under field conditions.

In one of the more graphic demonstrations of its ability to perform under the most trying conditions, two CAP planes were landed on a small stretch of gravel road one mile from ground zero on the day following the explosion. There in the shadow of a typical, two-story American home reduced to shambles by the explosion, the planes took on "survivors" and winged their way to Yucca airstrip and safety.

"How is it," General Beau was asked, "that the Civil Air Patrol was chosen to take part in the atomic tests?"

"It is our philosophy," he said, "and that of the Federal Civil Defense Administration that in time of actual attack the planes and crews of the Army, Navy, and the Air Force will be busy with the actual work of defending us against additional onslaughts. In this air age we must have an air arm on the homefront. The Civil Air Patrol is the only nationwide, disciplined, volunteer, civilian, flying organization presently in the process of being trained and equipped to do this work. Otherwise, any of the other splendid aviation groups, such as the Aircraft Owners and Pilots Association or the Flying Farmers might have been selected.

"It is well to point out also that when the Civil Air Patrol was made the civilian auxiliary of the United States Air Force by act of the Congress it was charged with the responsibility of providing this type of support to the American people.

"CAP is a service organization. It's only excuse for existence is to serve the people of the nation, the states and the local communities. This is a service we are providing."

Roscoe Goeke, radiological defense consultant to the FCDA, put it this way:

"For the first time, contemporary civilian lightplanes flown by non-professional civilian pilots (all members of the Navada Wing, CAP) took part in an atomic test. These planes and pilots were part of Civil

Effects Test Project 38.1 which was concerned with radiological defense monitoring techniques. The purpose of the test was to determine whether or not light aircraft flown by volunteer civilian pilots, manned by volunteer radiological defense monitoring personnel using portable FCDA radiation detection instruments now available would prove of practical value in the actual detection and measuring of radiation produced by fallout as a result of a nuclear explosion.

"Although the technical data collected as a result of this test is still being evaluated, I can at this time say that this type of aircraft operated by civilian volunteers proved to be a practical tool for the detection and measurement of radiation, and I feel that in cooperation with state and community Civil Defense organization the Civil Air Patrol can make extremely valuable contributions in this field."

Based on the findings in Operation Cue, civil defense officials, nuclear experts and CAP units in other parts of the nation began serious testing and training for the radiological monitoring role. One such program was instituted by the Oak Ridge National Laboratory, Tennessee. Called "Operation Egg Hunt", the program, supervised by Oak Ridge personnel, involved the use of controlled radiation sources secreted in areas surrounding Murrell Field, Morristown. Sources were selected that would represent no danger to persons, animals or vegetation and health physicists from Oak Ridge were responsible for handling them and directing ground crews.

Eighteen CAP aircraft under the command of Maj. Glen T. McIntyre, commander of Tennessee Group IV carried technical personnel from Oak Ridge equipped with simple detection gear of the type that would be available in any typical community CD organization.

At the conclusion of the mission, Dr. K. Z. Morgan, director of the Health Physics Division at Oak Ridge, said:

"I have observed and I am convinced that air survey by organizations such as Civil Air Patrol is the only practical way of locating rapidly the large quantities of fallout material associated with military use of atomic weapons in time of war or with accidents involving large quantities of fallout radioactive material.

"During the past 12 years, I have been associated with problems of fallout and surface contamination from radioactive materials and I believe these problems can be resolved adequately in time of peace and war if civilian groups such as the Civil Air Patrol are adequately trained and equipped to locate and appraise the hazard.

"For the past several years our division has had a program of air survey to map out background radiation from natural sources—minerals containing uranium and thorium. This program, in cooperation with the geological survey, has proven that it is very practical to use properly equipped aircraft to locate naturally occurring sources of uranium and thorium and it is my conviction that we should train personnel to use aircraft and properly designed equipment to locate fallout material that might be expected in case of enemy action.

"Radiation is no more dangerous than any other normal hazard.

The only difference is that it requires special instruments for detection. If these instruments are airborne by properly trained Civil Air Patrol personnel, one man can survey in a matter of hours what would require days or weeks to cover with surface borne equipment.

"I believe that one of our best sources of national defense against fallout material associated with the military use of atomic weapons in time of war is an adequately equipped and properly trained Civil Air Patrol."

Programs like Cue and Egghunt were of major importance to the Civil Air Patrol. Immediately after the conclusion of World War II mission activity dropped off significantly as was expected. But what was not expected was the apathy that gripped the nation and the almost negative interest shown in anything that was defense oriented. Certainly the professional military man understood. What was occurring was no different than what occurred after every sustained period of major conflict. But the members of the CAP were not professional military men. They were just everyday citizens—perhaps with special skills and a special interest—but on the whole very ordinary. Where only a few months or a year or two before they were operating at peak efficiency and found public and governmental support at virtually every turn, they now found themselves literally forgotten men and women. Very much like the CAP members of today, they could get along without the recognition but not without their life blood—the challenge of the mission and satisfaction of accomplishment. The realization by Civil Defense authorities that CAP and its lightplanes could play a major role in their program proved to be a shot in the arm and citing the proof positive of Operation Cue, CAP wings began a general assault on their individual state houses. In a few instances they were welcomed with open arms and CAP/CD state agreements were quickly forthcoming. In many states, however, CD authorities were reluctant to share their mission with anyone else. Human nature being what it is, this could have been expected. CD organizations also had enjoyed a long period of activity and now were faced with little to do. In these states, CAP found it tough sledding despite the fact it was patently apparent that CD required an air arm and CAP logically was it.

A real breakthrough didn't come about until early 1961 when the CAP National Board, Air Force officials and Federal Civil Defense authorities reached an agreement that resulted in a National Emergency Mission for Civil Air Patrol.

The implementing mechanism was a CAP-CD plan which consisted of a model agreement to be used by regions and wings as guidance in establishing uniformity throughout CAP in setting up working agreements with state Civil Defense offices.

Under the CAP-CD plan, Wing commanders were directed by CAP Wings and state Civil Defense agencies. Those wings that already had working agreements in effect were asked to re-examine their agreements with a view toward effecting uniformity throughout CAP.

The model agreement outlined approved duties and responsibilities of CAP/CD effort adequate to deal with disaster resulting from an attack and to provide for adequate civil defense.

The manner in which CAP personnel, property, and equipment would be utilized as organized units under Civil Defense would be determined by the commander or acting commander of the CAP wing concerned.

All CAP participation by a wing in Civil Defense operations would conform with the particular state's Operational Survival Plan in order to insure the most effective utilization of the volunteered manpower and other resources. It would also conform with regulations and policies as set forth by the Civil Air Patrol corporation.

CAP units would normally be prepared to perform aerial radiological monitoring, courier and messenger service, aerial surveillance of surface traffic, light transport flights for emergency personnel and supplies, aerial photographic and reconnaissance flights, radio communications, and other services within the capabilities of CAP.

Considerable headway was made during the 1960s with CAP gaining more and more recognition at the state level for its unique capabilities and with that recognition came acceptance. However, as in all other areas where a national entity must be integrated with a state or local agency, there continued to be roadblocks and obstacles in many states. Another major step in the right direction came in 1970 when the FAA, the Department of Transportation and CAP signed a memorandum of understanding designed to "enhance the maximum effective use of non-air carrier aircraft during time of national emergency." Under the memorandum, the FAA recommended that CAP be incorporated into State and Regional Defense Airlift Plans (SARDA). SARDA plans outline the procedures for mobilization of all aircraft and supporting resources within the state—other than those assigned to the Federal operations (Civil Reserve Air Fleet-CRAF, and the War Service Programs-WSP.)

FAA authorities also pointed out that "the Civil Air Patrol represents a unique opportunity for the Air Force to assign specific missions and tasks to a segment of non-air carrier aviation in support of its war plans. In doing so, it emphasized the Air Force "should be assured of a priority response on those missions it considers vital to the defense of the nation".

In transmitting its recommendations to the states, the FAA allowed for "variations to the suggested handling of SARDA/CAP organizational relationships "to accommodate local or state requirements or to be compatible with other agreements and plans having an influence on the CAP or SARDA."

The plan recognized the requirement for "state support" of CAP units conducting missions for the USAF stating that "units involved remain a civil resource even though the CAP is an auxiliary of the USAF and their tasks may be military in nature.

"Their aircraft will require civil fuel, operate from civil airports, and will need maintenance from civil repair agencies. Accordingly, resource managers should plan to allocate state resources for all CAP operations, both civil and military, from the same sources that will sustain SARDA."

A key element of the FAA SARDA proposal for state implementation recognized certain capabilities of CAP "not apparent in other general aviation resources" that make up a SARDA organization.

"Additional specialized skills," the FAA declared, "must be acquired in specialities of mission coordinators, clearance officers, ground operations officers, and communications officers, for example, if SARDA is to be successful. These special skills are available to the states in CAP units, but not in sufficient numbers to accommodate an expanded emergency airlift or without the possibility of jeopardizing the effectiveness of the Civil Air Patrol. Although some cross utilization of the CAP skills with SARDA may be practical, in some instances, many voids will remain."

To overcome this deficiency, the FAA concluded, "it is recommended that state aviation (SARDA) officials arrange for the training of key SARDA personnel through the CAP."

A final recommendation in the plan agreed upon by FAA administrator John H. Shaffer, CAP National Commander Brig. Gen. Richard N. Ellis, USAF, and CAP National Board Chairman Brig. Gen. F. Ward Reilly, CAP left little doubt as to the status and overall capability of the CAP resource insofar as it was recognized at the Federal vantage point. It stated:

"State SARDA and CAP officials are encouraged to enter into formal arrangements to enhance the effective use of state aviation resources in time of national emergency. Such arrangements may include assignments of the CAP wing and subordinate CAP organizations to: (a) serve as a primary emergency operational staff for the State Director of Aviation and other SARDA officials at satellite airports, (b) provide emergency services training for non-CAP personnel, and (c) perform specific emergency services, including those in support of USAF war plans. Arrangements or agreements between Civil Defense agencies and the Civil Air Patrol should also be reviewed, where necessary, to provide the means for rapid response to Civil Defense needs."

These concepts formulated at the seat of government for utilization of the Civil Air Patrol in time of national emergency or natural disaster have, over the years, been well thought out and are based on careful and thorough evaluation of CAP's demonstrated capabilities—in wartime missions, in air search and rescue and in providing unique services in time of fire, flood and storm. Yet, today it would be grossly inaccurate to report that in all 50 states, the District of Columbia and the Commonwealth of Purerto Rico CAP resources are recognized and accepted at face value, particularly to the degree that existing national agreements would have it.

True, in many states the Civil Air Patrol is more closely integrated with state CD/disaster agencies than even the architects of the national plans envisioned. In other states, however, this is not the case. Here the basic, long-standing political concern over Federal direction versus so-called state's rights comes into play with the same vengeance displayed in the relationship between full-time, official Federal and state agencies. It is the same disturbing force that has to a significant extent prevented the full acceptance of uniform national standards in areas such as law enforcement operations, building codes, professional licensing and environmental impact to name a few.

Fortunately, the states in which parochial considerations continue to impede full utilization of CAP capability and resources are in the minority. And, perhaps, in the broadest sense, the outstanding degree with which CAP has been accepted into the state family in other areas more than makes up for those in which optimum cooperation is lacking.

Typical of the states where the Civil Air Patrol wing and its resources are clearly identified in state civil defense disaster relief plans are Illinois, Hawaii, California and North Carolina.

In the nation's island state, for instance, the CAP's Hawaii Wing is fully integrated into the CD structure. Under normal circumstances command and control is vested in the wing commander, who, in turn, receives the CD mission requirements and determines the priorities with reference to any on-going USAF requirement. The Hawaiian Plan, however, also provides for those situations when a CAP unit may be isolated and without communications to its command echelon. In such situations, the CAP unit responds directly to the tactical control of the local CD agency (or other government agency if required) and continues to function in support of that agency until normal command channels are restored.

Under the Hawaiian CAP/DC agreement, the Civil Air Patrol has its own facilities in the State Emergency Operations Center (EOC)—a huge, protected tunnel in the Oahu mountainside and, in addition, has its own EOC in a nearby tunnel. The islands are divided into Rural Area Commands (RAC) with task forces assigned to support each RAC. CAP units throughout the islands are integral elements of these task groups.

Missions called out for Hawaiian CAP units under the plan are divided into three main groups—air operations, communications and support. Air operations for which wing personnel are equipped and trained include search, rescue and life-saving or life-supporting flights; aerial reconnaissance/damage survey/photo flights; airborne communications relay in support of tactical operations; airlift of critical supplies or personnel; other flights as required to support either CD or CAP tactical missions; and flights for general public welfare including airlift for news media when properly authorized.

Communications missions called out are those supporting tactical air missions with emphasis on air operations priorities; those involved in relaying radiological data, damage assessment information or other essential intelligence; supporting other tactical missions of CAP, the state, county or local CD authorities and administrative communications in support of either CAP or CD units. CAP personnel not involved in either air or communications operations per se are tasked to provide essential tactical mission support—transportation, logistics, supply, medical, EOS support such as shelter status and resupply; administrative support and special support for CD such as messenger and plotter duties.

Hawaii's comprehensive plan also provides for CAP aircraft and crews to function under SARDA either in time of military emergency or natural disaster provided a "Civil Defense Emergency" is declared by the governor. SARDA operations envision CAP providing communications between "control airports" and SARDA headquarters as well as providing "self-contained mobile support air units" (which would correspond roughly to wing and squadron operations sections and their equipment. Under the plan, CAP also could be asked to supervise private, non-CAP aircraft and crews in the mobile support air units.

A further measure of the thoroughness of the Hawaiian operations plan for CAP in time of emergency can be found in the annex devoted to Security Control of Air Traffic and Navigation Aids (SCATANA). Hawaiian CAP and CD authorities have left little to chance. The SCATANA annex clearly sets forth the manner in which each CAP aircraft will be controlled during a war emergency, the priorities of air movement and individual dispersal assignments. Dispersal airfields are called out and aircraft by numbers, types and home base are assigned to each dispersal field.

Hawaii represents a state where the Civil Air Patrol forces are integrated directly into the state CD/disaster relief organization with the wing commander directly responsive to the state CD director (except when engaged in performing missions for its parent service, the Air Force). Illinois on the other hand is one of the many states where CAP—in its disaster relief/CD role—functions under the direction of the State Department of Aeronautics.

In Illinois, the Department of Aeronautics is empowered by law to "supervise and to coordinate within and for the State of Illinois emergency or civil defense measures relating to all general aviation aircraft (other than CRAF) made necessary by local or national emergencies, including natural disasters."

The CAP/State CD/Illinois Department of Aeronautics agreement is most clear in both intent and purpose where CAP is concerned. It states categorically:

"During a civil defense emergency declared by the governor or the director of the Illinois Civil Defense Agency, the Department of Aeronautics will employ the Illinois Wing, Civil Air Patrol, using its facilities, personnel and equipment to support the Illinois Civil Defense Agency together with its other missions as a volunteer civilian auxiliary of the United States Air Force."

CAP's missions under this agreement also are clearly stated as aerial radiological monitoring, courier and messenger service, aerial surveillance of surface traffic, light transport flights for emergency personnel and supplies, aerial photographic and reconnaissance flights, radio communications and other services "within the capabilities" of the Civil Air Patrol. The agreement provides that subordinate units of CAP located within the jurisdiction of a county or city CD agency will participate in the local CD organization as a unit unless specifically assigned other duties by the wing commander. CAP units so deployed serve under their own unit commander who, in turn, is responsive to the local director of CD.

While virtually all CAP/state agreements negotiated prior to the mid-1970 period emphasize the civil defense role (the action to be taken in the wake of a nuclear attack), these now are being gradually modified to place emphasis on responding to natural disaster. New agreements coming into being are, in general, following the philosophical lead of the Department of Defense, which in recent years inherited the so-called civil defense responsibility at the Federal level.

At first the Defense organization established to pick up where the old Federal Civil Defense Agency left off was known as the Office of Civil Defense (OCD). Now, however, it has been designated the Defense Civil Preparedness Agency (DCPA) and as such its responsibilities have been broadened to encompass not only pure civil defense (wartime) considerations but also to include planning on a national scale to meet the requirements of natural disaster.

This is not to say that the nation is letting down its guard with respect to a possible nuclear attack, but rather is modifying its position within the context of the detente existing between this country and the Soviet Union, the new spirit of cooperation between the United States and the People's Republic of China and the continuing encouragement coming out of the Strategic Arms Limitation Talks (SALT).

In the minds of some CAP old-timers, this changing philosophy appears to bring with it a "weakening" of the CAP mission. They feel the sense of urgency that accompanied the military preparedness aspect of civil defense is lacking.

Circa-1970 CAP leaders feel just the opposite—and they are in the large majority. They see a greater need for diversified training, more opportunity to employ CAP's unique airlift communications capability and, above all, much closer ties to local law enforcement/disaster relief organizations with more opportunity for CAP's resources to be called upon to assist in local emergencies.

Although with the SALT agreement and the detente the United States enjoys today with both the Soviet Union and the People's Republic of China the possibility of nuclear attack is significantly lessened, the nation's civil defense guard must remain up. CAP remains a major resource for CD. Here decontamination team member hoses down CAP aircraft which has penetrated simulated

CHAPTER 9

CAP And The "Bug"

It was mid-Spring 1957. A group of scientists, an Air Force colonel and a Civil Air Patrol officer huddled around a conference table in Washington, D.C.

The entrances to the building were guarded by vigilant sentries. Individuals entering the building were challenged and without the proper identification were turned away.

Under discussion was the United States Earth Satellite Program—a part of this nation's participation in the International Geophysical Year (IGY). As unlikely as it seems today, (it also appeared as unlikely then to CAP Col. Donald T. Spiers, commander of CAP's Middle East Region) the scientific community and the Air Force were asking for CAP's help with the nation's embryonic space effort.

Specifically, they were concerned with the methods then envisioned to keep track of the tiny, 20-inch, metal ball jammed with electronic equipment that was planned to be hurled into earth orbit by the Navy's Vanguard rocket. More directly they were concerned with how to train hundreds of volunteer astronomers—amateurs as well as professionals recruited by the Smithsonian Astrophysical Observatory to provide the manpower for the Smithsonian's Satellite Optical Tracking Program.

Involved with Colonel Spiers in that initial meeting were Dr. Armand N. Spitz, coordinator of the Visual Operations Portion of the Earth Satellite Program; Dr. J. Allen Hynek, associate director in charge of the Satellite Optical Tracking Program; Dr. Fred J. Whipple, director of the observatory; Col. Owen F. Clarke, assigned by Headquarters, USAF, to assist in the program.

Their major concern was how to keep track of the satellite after it entered its orbit around the earth. With the ever present possibility that the relatively few electronic tracking stations and astronomical observatories would lose the little speck of light—about equal to a fifth or sixth magnitude star—it became apparent that many eyes would be necessary. To correct the situation, a program known as Operation MOONWATCH was born.

The Smithsonian Astrophysical Observatory agreed to undertake the task of obtaining hundreds of volunteer astronomers across the country. Leon Campbell was named supervisor of operations for the program and in a relatively short time enough volunteers were recruited to man more than 100 Operation MOONWATCH observation posts.

A special low-cost MOONWATCH telescope was developed. Some of the volunteers built their own with funds from their own pockets and some with funds furnished by community organizations. The volunteer moonwatchers set out to equip their stations for the big day.

At this point, however, the men guiding the optical observation part of the satellite tracking program faced a new problem—how to provide training for these scores of volunteers.

Hynek turned to the Air Force. It had been computed that if a light of a specific magnitude were towed behind a jet fighter flying at an exact Mach number at 35,000 feet it would appear in the moonwatcher scopes very much like the first earth satellite was expected to appear. It would, however, take thousands of flying hours and hundreds of aircraft to accomplish the job. For both operational and budgetary reasons the Air Force was forced to turn the scientists down—as far as jet fighters were concerned. Colonel Clarke had suggested an alternative. Why not use the Air Force Auxiliary, the Civil Air Patrol, which had thousands of pilots and hundreds of airplanes— airplanes that could be flown for pennies when compared with high performance USAF aircraft?

Contacts with CAP Headquarters found a "can do" attitude and in a matter of weeks a full-blown program had been designed. Colonel Clarke would provide Air Force liaison between the Smithsonian and the CAP. Colonel Spiers was named project officer for the corporation and the Anacostia Squadron, National Capitol Wing (then commanded by the author) would perform the initial research and development tests.

Air Force technicians got their heads together with their counterparts at the Naval Research Laboratory and determined what reductions in altitude, airspeed and magnitude of the light were necessary to bring the pseudo satellite down from jet altitudes and speeds to those compatible with CAP aircraft. The results: a standard 1.5-volt flashlight bulb powered by a pair of "D" batteries, moving at 105 knots at exactly 7,000 feet above the terrain where the MOONWATCH station was located. Now that the scientific computations were completed it took a little old fashioned "Yankee ingenuity" to develop a training device. How to combine the batteries, socket for the bulb and a dropping-resistor to limit the light to one-tenth candlepower into a package that could be pulled in a stable mode behind an airplane?

The familiar "plumber's helper"—the rubber bulb used to clear a balky drain—was the answer. The wooden handle was detached and a metal ring on a swivel inserted in its stead. A simple metal bracket was fabricated to attach to a swivel inside the cone. The bracket held the batteries in a standard clip and provided a mounting extension for the bulb socket protruding about three inches from the rear of the rubber cone affording maximum visibility. The entire arrangement, soon dubbed "the bug" by CAP crews across the nation, was towed behind the aircraft using 100 feet of plastic covered clothesline. The line provided the necessary visual separation between the target light and the aircraft's navigation lights. The rubber plumber's helper served two purposes. First, it was the "perfect aerodynamic shape" to stabilize the device in tow. Second, in the event it accidentally became detached from the tow line, the shape also would stabilize on descent and the flexible rubber cup would protect anyone on the ground from serious injury.

For several weeks, the Convair L-5 of the Anacostia Squadron performed flight tests with the bug and pilots practiced the mission profile working toward a high degree of precision with regard to airspeed, course and altitude. Full scale operational testing would be accomplished in cooperation with MOONWATCH stations established at Springfield and Fort Belvoir, Va., near Washington.

The course to be flown by the L-5, over and over again until only enough fuel remained to return to Hyde Field in nearby Maryland, lay directly in the area where several major airways converged. The MOONWATCH station location coincided almost exactly with a major in-bound reporting point for aircraft approaching from the south and west landing at Washington National Airport. The L-5 was equipped with dual channel, HF radio with one of the channels set up for CAP air/ground communications with the MOONWATCH stations. The night was black, black and moonless. The periods when satellite visual tracking would be performed—the first few hours after dark and before dawn are high traffic periods for Washington National. In short, the situation was far from optimal.

The L-5 liaison plane cruised at exactly 7,000 feet over the Virginia countryside. The crew was tense. Twisting around in the cramped confines of the rear seat, the observer, Senior Member Roy Homer, began playing out the plastic-covered braided-wire clothesline coiled behind the seat. As the last of the line was played out and the slipstream drew it tight, the radio crackled.

"YOU ARE OVER THE SPRINGFIELD BEACON. YOUR NEW HEADING IS ONE TWO SIX DEGREES. WE WILL GIVE YOU THE NECESSARY CORRECTIONS TO KEEP YOU ON THIS TRACK. WHEN YOU PASS OVER FORT BELVOIR, MAKE A PROCEDURE TURN TO THE RIGHT AND RETURN ON THE RECIPROCAL OF THIS HEADING. WASHINGTON CENTER WILL PICK YOU UP OVER THE SPRINGFIELD BEACON AND PUT YOU ON TRACK FOR YOUR NEXT PASS."

As the small CAP plane began its track, more than a score of astronomers and scientists manning the Fort Belvoir, Va., and Springfield MOONWATCH stations gazed into their scopes each one mounted and calibrated so as to view one small segment of a 120-degree slice of sky along the 78th meridian.

Overhead the engine of the plane could be heard. Those not watching through scopes could see its navigation lights. The questions in every mind were: Will the tiny light do the job? Will it appear in the scopes as it should? Will the CAP plane and its crew of non-professional volunteers be able to perform the precision flying necessary to insure that the speed, course, and altitude will be exactly correct?

Suddenly a man down the line of scopes called, "Number 7!"

At the same time he pressed a button which activated an electronic timer. Then another and still another sighting was reported.

Colonel Clarke, the AF project officer for Operation MOONWATCH, and Doctor Spitz, on hand for the experiment, breathed sighs of relief. The tests were going to be a success.

Two hours and many sightings later, both Colonel Clarke and Doctor Spitz expressed their "entire satisfaction" with the tests.

With the real earth satellite being rushed to completion for launching during the International Geophysical Year, the scientists had been hard-pressed to provide advance training for the more than 4,000 volunteer observers. Now, thanks to the CAP, these observers would have many hours of actual practice in sighting, tracking and reporting on the movements of the satellite—or any satellite for that matter—long before its epoch-making voyage into space.

In a matter of days, the Air Force began putting the loose ends together. One hundred of the "bugs" were ordered constructed for distribution to CAP units situated near MOONWATCH posts all across the country.

An official request was made of Maj. Gen. Walter R. Agee, CAP national commander, for full scale support to the hundreds of volunteer astronomers and scientists to be coordinated by the Smithsonian Astrophysical Observatory.

Within a few days of the first full-scale operational tests of the bug and of CAP's capability to perform the MOONWATCH mission, a hundred "simulated satellites" were under production destined for shipment to CAP units across the nation situated geographically where they could support the MOONWATCH training program. Procedures for flying the mission, safety precautions and instructions for obtaining USAF reimbursement for fuel and lubricants for the aircraft were being prepared at CAP headquarters for distribution to the field.

A few nights after the initial successful test, CAP was called upon to conduct a demonstration for a large group of military and civilian dignitaries. The Anacostia Squadron L-5 was airborne again, Roy Homer in the observer's seat handling the bug and the author at the controls. It was October 4, 1957.

The ground observers seemed never to get enough. Again and again Washington Center vectored the L-5 back to the starting point for another pass. "That was an excellent pass," the word came by CAP radio from the ground, "but they would like you to do it again." In the "G" model L-5 being flown it was necessary to leave the ambulance section door open when the bug was deployed. Near freezing air blasts at 7,000 feet were chilling the crew to the bone. Finally, since fuel was running low, they radioed a request to return to base. The radio response was electrifying:

"AERONAUT TWO YOU MIGHT AS WELL GO HOME. THERE IS A REAL SATELLITE THERE NOW!"

The Soviet Union had launched Sputnik One. The space age had begun in earnest.

Russia's historic feat, the successful orbit of the world's first artificial satellite, put Operation MOONWATCH on a "crash" basis. The following day, Colonel Clarke asked General Agee and the Civil Air Patrol to accelerate its program. The Air Force authorized CAP to expend an initial 10,000 flying hours on MOONWATCH—a million miles of flying over the next year.

Colonel Clarke urged CAP commanders the nation over "to immediately put your MOONWATCH program into effect."

Detailed instructions and a standing operating procedure for all CAP MOONWATCH flight operations were rushed into print and dispatched from CAP National Headquarters.

Some 31 CAP Wings and squadrons already had the simulated satellite designed by the USAF and Navy researchers for use in the training program.

"Attempts to observe the Russian-built satellite," declared Colonel Clarke, "have emphasized the need for training on the part of the hundreds of volunteer astonomers, manning the 90-odd MOONWATCH posts in this country. This target towing assignment given the Civil Air Patrol is designed to provide that training.

"I urge all CAP commanders to implement immediately their MOONWATCH program in accordance with the directives provided by higher headquarters. In fact, because we are behind in starting this important program, I recommend that CAP commanders and MOONWATCH mission commanders contact the director of their MOONWATCH post at once and apprise them of CAP's readiness to begin the tracking missions. It also is desirable that more missions than originally scheduled for each station be performed."

Authority to approve requests for MOONWATCH missions was delegated to the respective wing AF-CAP liaison officers. Instructions for applying for reimbursement also were rushed to the liaison officers.

While the liaison officers were responsible for controlling as well as approving the tracking missions, they were authorized to use "on-the-scene" CAP mission commanders to assist in the exercise of control. Liaison officers were instructed not to authorize a mission when the aircraft or the crew or any other factor involving flying safety did not meet the requirements set forth in a published standing operating procedure published by National Headquarters.

In addition to the MOONWATCH aircraft, three CAP mobile radio cars were authorized reimbursement for each mission.

Operation MOONWATCH taxed CAP unit resources. In those days only a relatively few corporate aircraft and a limited number of member-owned planes were equipped with the radio gear, instruments and night flying equipment necessary in the interests of safety for this mission. Of some 5,000 available airplanes, only a few hundred could be used. It also was necessary to insure that all pilots were night qualified and use of a second rated pilot to operate the bug was recommended.

CAP's performance of this little known but vital mission was on a par with its performance 15 years earlier when it was called upon to meet the U-boat threat. In words of both Air Force and Smithsonian officials, it was "outstanding." It, however, wasn't the Civil Air Patrol's only contribution to the U.S. space program. Within weeks of the launch of Sputnik One and the real beginning of Operation MOONWATCH, the Society of Photographic Scientists and Engineers requested CAP help again. This time it wasn't the pilots and observers who were needed, it was the thousands of CAP radio operators.

Norton Goodwin, secretary-treasurer of the association told CAP that the world's scientists didn't actually know exactly what shape the earth was—that is to the finite degree necessary, for instance, to plot the trajectory for an intercontinental ballistic missile or a very precise space launch vehicle.

According to Goodwin, the Society of Photographic Scientists and Engineers had established a "phototrack" program on a nationwide basis which ultimately would supply such information.

Goodwin emphasized that at that point in time cartography the world over, in the Soviet Union as well as in the United States was based on a series of educated guesses as to the exact configuration of the earth.

"It is true," he said, "that we know the earth actually is an ellipsoid rather than a sphere, but so far there has been no means for accurately determining the exact shape of this ellipsoid."

The phototrack technique to be used by the society in connection with the International Geophysical Year program— the satellite program of the IGY to be specific—was expected to provide the means for "fantastically accurate" measurement of the earth's surface as well as information as to the shape and effect of its gravitational field.

Dr. Francis J. Heyden, director of the Georgetown University Observatory, said that this information would be "worth its weight in platinum, now!"

The society, with some 800 regular members all highly trained, asked each member to maintain a phototrack station. These stations were situated at a United States Coast and Geodetic Survey Triangulation Station or certified offset. There are more than 100,000 USCGS Triangulation stations in the United States.

The Society of Photographic Scientists and Engineers appealed to the Civil Air Patrol for aid in flashing tracking data to its phototrack stations on a daily basis.

The program worked this way. The Smithsonian Astrophysical Observatory telephoned the tracking information to Civil Air Patrol National Headquarters in Washington each day. This information indicated whether a satellite would pass over the United States and whether or not conditions would permit its track to be photographed.

A typical forecast read this way:

"THE FOLLOWING ARE SATELLITE PHOTOTRACK ACQUISITION FORECASTS FOR THE PERIOD SIX THROUGH SEVEN NOVEMBER. THE CONDITION WILL BE BLACK AND WHITE."

The message indicated that all satellite passes over the United States during November 6 and 7 would be at night (black) or during the day (white) and there would be no visible passes that could be photographed.

Where green (visible because of reflected light) or red (visible because the object is self-luminous) passes were predicted, the data was given with reference to the path of the object as it intersected the 40th parallel of latitude:

"14 NOV 57A1 GREEN 0326 WEST 023 DEGREES 0536 EST 051 HIGH."

Such a message predicted that on November 14, earth satellite 57-A-1 would make a visible pass across the United States crossing the 40th parallel at 5:36 A.M. at the subpoint 82.6 degrees West of Greenwich on a heading of 023 degrees and at an altitude of 510 miles. This information could be marked on a transparent overlay on a 1956 National Geographic Society map of the United States and there converted to local azimuth and elevation information through use of the National Geographic Society's "satellite finder."

Each satellite was given a code designation. For instance, the rocket part of Sputnik One was known as 57-A-1; the satellite itself as 57-A-2. The forecasts were broadcast over the CAP National Headquarters radio station in Washington VICTOR PAPA ZERO at 1900, 2200 and 2100 hours daily. They were re-broadcast over CAP wing net frequencies during that portion of the hour after each national broadcast.

Phototrack stations either were equipped with receivers capable of use on the wing frequencies or, in many cases, CAP communicators brought their own equipment and became part of the phototrack team. As in Operation MOONWATCH, CAP volunteers ultimately were credited with a "major assist" in the successful accomplishment of the program.

Another short, but history-making chapter, was added to the annals of Gill Robb Wilson's brainchild.

The author in the Anacostia Squadron's L-5 gets a pre-mission briefing from a fellow CAP officer prior to one of the initial research development flights with the "bug". The device (shown in the inset) consisted of a familiar "plumber's helper" equipped with a metal support structure, pair of "C" batteries, dropping resistor, socket and lamp and was towed behind the L-5 on 100 feet of plastic-covered, metal clothes line. At exactly 7,000 feet AGL and at 105 miles per hour it presented the same visual target as the first earth satellites. Operations MOONWATCH and SPACETRACK were important CAP missions in support of early U.S. satellite programs.

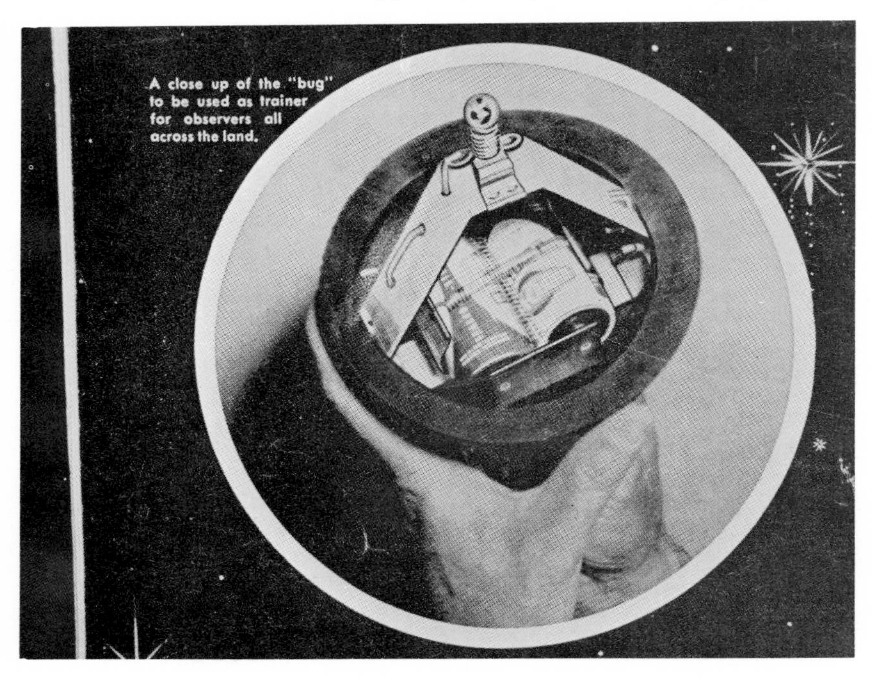

CHAPTER 10

CAP Spells "Help!"

It was 8:35 P.M., on a hot, sultry June evening in Flint, Michigan. Laughing young couples walked along Coldwater Road for a breath of air. Moviegoers who had just left the early shows were driving home. A tiny tot stood near the door of an ice cream store, crying lustily over his cone which was splattered on the sidewalk. It was just an average June evening in an average American town.

Five minutes later Coldwater Road was a nightmare of twisted wreckage, smashed automobiles, ruined buildings and broken bodies. The stillness that follows sudden death was punctured by the cries of the injured and the distant wails of sirens from downtown Flint signifying rescue already on the way to the scene.

The stumps of once proud young trees reached skyward like grotesque hands trying to hold on to life. Like a huge bowling ball, the tornado which struck Flint on this peaceful Monday evening rolled for eight miles down Coldwater Road smashing everything in its path. In Flint alone 113 persons were dead when the ominous black funnel had gone its way. In all, six tornadoes struck the midwest that day inflicting damage estimated at many millions of dollars. The area was declared a disaster area by the President.

When the Flint disaster was over, those who survived the death dealing twister began trying to rebuild their homes and businesses and even their lives.

Less than a half hour after the tornado struck, members of the Civil Air Patrol's Flint Squadron were in the midst of the disaster area, digging through the ruins for survivors.

Their portable generators were rushed to the scene to provide power for floodlights to light the rescue operations. They walked guard over the ruins through the night until the National Guard arrived, the next day.

Within minutes after the Flint unit went into action, the word was out and CAP units all over the state were alerting their personnel. CAP squadrons from Kalamazoo, Austin Lake, Allegan, Lawton, Mattawan, Battle Creek, Grand Rapids, Muskegon, Lansing, Flushing and other parts of the state were headed for the area by dawn Tuesday.

On their arrival they found that Brig. Gen. Lester J. Maitland, Michigan director of Civil Defense, had jobs for them all. As they arrived, radio-equipped cars were stationed at different locations in the disaster area, while others remained outside the devastated regions to relay information and requests for help to the outside world.

When the search and rescue bus of the Kalamazoo Squadron arrived with its mobile communications and telephone equipment, it became the headquarters for CAP operations in the disaster area and command of the entire CAP rescue effort was turned over to Maj. Robert C. Hamill, Kalamazoo commander. In addition to the radio cars, other CAP members with walkie-talkie radios roamed the area dispatching calls for additional aid and reports on the disaster.

This is the way one radio newscaster at the scene described the work of a group of CAP cadets:

"On the way to the stricken area, the cadets, some of them only 15 years old, acted as you would expect any 15-year-old to act—they played guessing games and munched gum. They filled up with pop every time we made a comfort stop.

"But when we hit the disaster area, the serious nature of the mission hit the boys. Actually they were no longer boys—they were men—doing men's jobs.

"In a matter of minutes, the basement and first floor of a gutted school building were cleared of rubble and readied for emergency shelter. While one of the cadets cleared the building, another group was stationed as guards around the building and throughout the disaster area. The cadets worked beside state troopers directing traffic. They patrolled beside armed National Guardsmen. Through their efforts looting was stopped. Thanks to their efficient searching, many valuable items were reclaimed and registered in their owner's names. CAP teams patrolled the ruins, checking all persons for proper identification. Those found without proper identification were turned over to the state police for investigation.

"You people at home can be proud of the men who are representing Southwestern Michigan in the heart of the worst disaster to hit the State."

Another radio reporter told his listeners:

"Civil Air Patrol deserves a great deal of commendation. Remember the boys who marched in the Memorial Day parade a week and a half ago—well, today they are the men who are doing a magnificent job of directing traffic, standing guard and participating in operation 'Cleanup'."

In a letter to the Kalamazoo commander, Robert Paul Dye, of the Fetzer Broadcasting Company, said:

"Your fine work at Flint also has given the people of Southwestern Michigan a feeling of security. They know now that in the event of a disaster in this area, the Civil Air Patrol can mobilize within a number of minutes. They know that their splendid system of communications can summon additional aid and keep contact with the outside world. I am not easily impressed. However, the work of your group impressed me to the extent that in cooperation with the other members of the radio and television staff, I am preparing a special show dedicated to the work and training of your group."

Ray Guiles of the Detroit Times wrote:

"The defense training of the Civil Air Patrol has paid dividends during the tornado emergency."

From the Pontiac Daily Press came these words:

"Pontiac's CAP group has had its baptism of fire. The local unit faced emergency conditions for the first time in the Flint tornado rescue operations and its men came through with great credit to themselves and to the Civil Air Patrol."

R. L. Faust, administrator of the Port Huron Hospital, cited CAP aid in furnishing emergency power with their portable generators when falling trees cut off all electric power to the hospital.

The Kalamazoo Gazette carried these words under the byline of Dan Ryan:

"The work of the Civil Air Patrol members from Kalamazoo and other cities in the state serves as a strong reminder of the value of preparation, organization and trained personnel in time of disaster."

The women of CAP also had their work cut out for them. Capt. Catherine L. Bush, director of Women's Activities for the Flint Squadron, and her staff relieved state police of the clerical detail of issuing passes to householders who wanted to probe the ruins of their personal belongings.

CAP units from Coldwater, Grand Rapids, Freemont, Albion, Bad Axe and Charlotte also aided in disaster relief efforts. In all, 16 Michigan units turned out when the state-wide alert was called.

To the plaudits of a grateful people, CAP's National commander, Maj. Gen. Lucas V. Beau, added his commendation for the job CAP members had performed.

Later in the same year havoc became commonplace in western Maine as rampaging rivers flooded farms, cities and highways. Communications outside flood areas were cut off. Homes were left to the mercy of the surging floodwaters and transportation in some areas was at a standstill. Following heavy rains, rivers began to rise above the flood level stages. Surging over dams and banks, the waters engulfed farm lands and homes. In the Rumford-Mexico area some 1,200 families had to abandon their homes March 27. More than 50 families were forced to evacuate their homes in Saco March 29. Forty families on the island between Saco and the twin cities prepared to leave their homes the same day as flood waters continued to rise.

As the flood threat increased, members of the Civil Air Patrol's Maine Wing went into action. CAP radio operations restored communications to the flooded areas. They aided flood victims, supplied small generators to replace malfunctioning electrical equipment and stood by to fly emergency evacuation missions. Through reports turned in by CAP aerial observers, disaster operations officials were kept informed of the flood situation.

A CAP radio message from operators in Rumford—one of the hardest hit areas—reported that up to five feet of muddy water covered every road isolating the Mexico-Rumford area. Most of the evacuations were from Mexico area that day, and many more were anticipated in Rumford.

The report indicated that one resident reported water at his doorstep and rising to the level of the living room floor. The Oxford

Paper Co. mill was shut down as the flood dangers hit western Maine's industries.

For some time, residents in the path of the turbulent Adroscoggin River feared that a big boom holding thousands of the Oxford mill's logs would increase the devastation. CAP members took up an emergency watch with portable radios ready to flash word down stream. The boom finally broke but did not tear out bridges in its path as had been feared. While the flood waters were still slightly below the level of the bridges the logs slid under the constructions.

At Biddeford, another hard hit area, eight CAP mobile radio cars and a fixed station went on a round-the-clock schedule to supply communications to the stricken community.

Several CAP generators were placed in the Saco-Lowell shops and other localities to generate electricity when needed. The Portland Squadron launched three planes over the flooded Saco River on reconnaissance missions.

Residents of York aided by CAP volunteers battled desperately with sandbags as waters continued to rise. Hundreds abandoned their homes during the night as the swirling waters invaded their front rooms and threatened to continue to rise. The Portland Weather Bureau predicted still more rain. CAP remained at the ready.

When prolonged drought conditions up and down the Northern Atlantic coast paved the way for widespread forest fires which burst forth in heavily wooded sections of Massachusetts and Virginia, Civil Air Patrol aerial observation and mobile radio communications were enlisted to aid the local firefighters in both states, the CAP fliers were called upon in other states to be on the alert for small fires which could have turned huge areas of timberland into raging infernos.

Showers across the Northeastern United States brought an end to the fire threat before the crisis ended, but large areas in Massachusetts and Virginia were devastated by blazes that broke out and raged uncontrolled for days.

Six persons, all members of one family, perished near Richlands, Va., when flames attacked their mountain home while the father, Harrison Osborne, was absent battling the forest fire burning in Tazewell County. Members of the Tazewell Squadron responded to a request for help sounded by the Tazewell fire warden. Portable communications equipment was dispatched to the area along with fire patrol teams.

Capt. Paul R. Puckett, Tazewell squadron commander, a veteran of a previous year's fire threat, directed the operations in the heart of the fire area. High winds fed the blaze for two days and nights and CAP radios were kept busy calling crews and relief workers. Action by firefighters along with light rainfall brought the fire under control late on the third day, but CAP aerial observers remained on duty until all danger had passed. CAP also contributed assistance by providing needed transportation in and out of the fire area.

In Massachusetts, the Turners Falls, Fitchburg and Reserve CAP squadrons provided men and planes to combat fires raging out of control near Charlemont.

A total of four separate fires were reported in central Massachusetts with the biggest conflagration on Mt. Negrus. Civil Air Patrol airborne communications enabled observers to report the fire's progress to watchers in fire towers, and to CAP mobile units on the ground. Rev. Stephen Tucker, chaplain of the Turner Falls unit was one of the first CAP members to take to the air in the fire alert. CAP members volunteered by the dozens as the fire spread, threatening to destroy thousands of acres of valuable timber.

CAP units from Methuen put walkie-talkies into action, and a communications network was set up along the entire fire line.

Allen C. Fisher, Jr. is one of the nation's foremost magazine writers, for many years a staff editor for the National Geographic Magazine. He also is one of the few writers who has had the opportunity to see CAP in action from coast to coast as he, as the guest of the U. S. Air Force, toured CAP units from Maryland to California preparing a major Geographic article of the Air Force's "little brother".

Fisher's "tour of duty" with CAP ended with the writer expressing high praise for its members and their contributions. This excerpt from his final manuscript tells why:

"Stroudsburg and East Stroudsburg, Pennsylvania, made national headlines when rains from Hurricane Diane inundated resorts in the Pocono Mountains. But few of these stories related the heroic work of Stroudsburg's able Civil Air Patrol squadron.

"The flood struck at night. Brodhead Creek, separating the two communities by 100 yards, swelled into a raging torrent. At the height of the storm, Warrant Officer Philip Hardaker packed a portable radio on his back and crawled across a trembling railroad trestle, luckily reaching East Stroudsburg before the bridge collapsed.

"For the next 24 hours that small radio was East Stroudsburg's only voice.

"Meanwhile, in Stroudsburg, Capt. William A. Bechtel jumped into the squadron truck and evacuated dozens of people from threatened homes. Providentially, CAP had cached emergency food and medicine in the area, and the next day he dispatched squadron planes for these supplies.

"Armed services helicopters began arriving. Bechtel set up a heliport in a schoolyard and assigned CAP members to fly as observers. They pinpointed marooned families for pilots who made pickups and dropped supplies.

"Other volunteers relieved Bechtel and his men after they had been on their feet more than 48 hours. Many were nearly incoherent with fatigue; some of these workers had lost their own homes.

"So many bodies were found that CAP personnel helped store them in refrigerator trucks as temporary morgues. More than 80 persons lost their lives in the Stroudsburg area."

Two disasterous hurricanes dumped flood waters on vast areas of Pennsylvania and Connecticut that year. The governors of those states subsequently summed up for Fisher their impressions of CAP's role in disasterous relief efforts like this:

"Wherever I went, I found Col. Phillip F. Neuweiler's state Civil Air Patrol wing on the spot," wrote Governor George M. Leader of Pennsylvania. "Reports reaching me later backed up my own impression:

"The CAP had not only done outstanding rescue work at the height of the flood, but the wing had pitched in magnificently on the staggering job of rehabilitation."

Governor Abraham Ribicoff of Connecticut said:

"Town officials throughout the State have praised CAP's assistance in furnishing generating equipment, in sending and receiving urgent messages through its statewide radio network, in supplying field telephones in the disaster areas, and in flying 6,000 pounds of badly needed food and medicine into the stricken communities.

"In addition, CAP flew observation missions and reported to State Civil Defense headquarters on conditions that existed in the flooded areas."

Almost whenever and wherever disaster strikes you will find the blue-garbed cadets and senior members of the Civil Air Patrol in action. In some states it is virtually an annual occurrence since nature seems to single out certain areas more than others. The "hurricane belt" along the gulf and eastern seaboards is one of these.

Hurricane Donna's week-long orgy brought forth tragic, but glowing reports on round the clock vigil and toil by scores of Civil Air Patrol units and hundreds of senior and cadet members.

The emergency power and communications systems set up by CAP all the way from Florida to Maine starred in the drama. Civil Defense and other agencies—Air Force, Coast Guard, National Guard, Navy, Red Cross, local law offices, state conservation departments relied heavily on the CAP volunteers who were everywhere.

Florida was one of the hardest hit wings, and from Naples and other areas came appalling stories of devastation.

From the log of Naples, CAP Composite Squadron, southwest Florida:

"September 10, 1205 hours: Eye of hurricane Donna over Naples; no wind at all . . . for about 30 minutes . . . then the fury of the storm hit . . . winds of 192 mph . . . for 12 hours Naples was battered . . . U.S. Coast Guard helicopters saw a ghost town . . . " The day before, Lt. George Spencer, Naples squadron commander, had personnel move aircraft; one L-5 was stored in the City of Naples hangar; an L-5 was dismantled and stored in downtown Naples. Only the L-4 was operational after the storm.

The same night, the Naples unit installed communications in the Civil Defense Headquarters high school shelter, at radio station WNOG, the municipal hospital and at Squadron Hall, manning the stations throughout the storm and for several days after. This was virtually the only communication the city of Naples had during the ensuing storm.

At Naples Municipal Hospital, Maj. Vernon Sweet of Group 8 and Capt. Harold S. Glazer of the Naples Squadron kept the hospital generator going for emergencies. CAP restored communications at Civil Defense headquarters and a large generator replaced damaged ones at the telephone company.

Lt. George C. Spencer, Naples squadron commander flew with the Coast Guard in helicopters to survey the Naples area. He saw a ghost town. Here and there a sign of life but mostly—nothing but rubble, and water. Housetops peeled off; tile and shingles down to clean wood; squashed homes with furniture strewn everywhere; trees uprooted, others snapped off at the trunks; aircraft blown from landing strips; a big, new school ripped apart as though an ogre had torn it with his giant hands.

Immediately following Donna's Naples debacle, cadets manned CAP communications points on a 24-hour basis. It is estimated that senior members and cadets of the Naples CAP units manned communications for 395 hours, while others operated as rescue cleanup details and did air spotting of trouble areas.

The Central Florida Squadron furnished emergency electrical power to the Kissimmee Florida Hospital, the Central Blood Bank and Herndon's Ambulance Service. Later, the generators were used to provide power to a local dairy for milking their cows. Squadron personnel also flew Red Cross evaluation teams over the area.

Bruce M. Thogmartin, administrator of Kissimmee Hospital later wrote:

"I, the staff, physicians and ancillary personnel of the Kissimmee Hospital, would like to take this opportunity to express our deep felt thanks and gratitude for the tremendous help provided by your squadron during the rampages of Hurricane Donna. Without the help of your mobile power plant, I shudder to think what might have happened to our obstetrical nursing, emergency and operating departments.

"I would especially like to commend CWO Jerry Genaw, and his three crew members George Carl, Cadet 1st Lt.; Robert Slaten, Cadet 1st Lt.; and Marvin Jewell, Cadet 3rd class, for their outstanding performance during this crisis.

"The hospital organization as well as the patients will always have a warm feeling in their hearts when they recall the efficiency which the above crew demonstrated in accomplishing their mission. I cannot find proper words to express our gratitude. Thanks so much for a job well done."

Daytona Squadron, CAP, spent more than one thousand man hours and used three power units to assist Holly Hill and Ormond Beach, Florida. Food had been stored in community freezers and power units were taken from freezer to freezer to preserve the food.

In North Dade County, Florida, Civil Air Patrol assisted in the storing and later flying out 350 aircraft. Mobiles and other vehicles assisted in evacuation while squadron radios provided communications between public works, Key West and the Homestead Pumping Station in connection with a broken water main leading to Key West.

The two Key West planes flew 52 hours searching the sea locating survivors, bodies and property such as boats, house trailers, docks, furniture.

The Anna Maria Island Squadron of Florida Wing provided generators for the sheriff's department. Members carted fallen trees from highways and streets and assisted in spotting downed power lines. The squadron maintained watch at two standby ambulances.

Palm Beach, Winter Haven and Fort Pierce, Florida, squadrons maintained constant radio watch and Manatee County Search and Rescue squadron stepped in with communications when public utilities failed. The Vero Beach unit used their personal automobiles to transport civilian supplies to hurricane shelters.

Alabama Wing CAP personnel continuously monitored the Civil Air Patrol radio frequency through Saturday and Sunday relaying messages for Georgia and Florida Wings.

Georgia Wing became the big communications control center for much of the eastern seaboard activity, with Lt. Sue T. Pattillo manning Red Star 3 in Atlanta.

The Brunswick-St. Simons squadron with Capt. Hamlett and Lt. Lawrence S. Miller, Jr., on watch at Red Star 42, was like a "switchboard" for the southeast, relaying radio messages and citing weather conditions from Clearwater, Fernandina and Daytona, passing messages on to Civil Defense Headquarters in Atlanta. Red Cross priorities were relayed to Macon and Warner Robbins, Ga., and Civil Defense traffic was moved into Tennessee and to the Maryland Wing as the storm whipped eastward.

In South Carolina, units responding to the emergency brought forth 48 senior members, 32 vehicles, five mobile generators, 18 fixed radio stations and 30 mobile relief units.

A tornado or two were mixed in with the trouble already caused by Donna in North Carolina. More than 250 CAP members turned out with 50 vehicles, 40 mobile units, 26 fixed radio stations, 15 mobile generators and CAP aircraft.

In the Kinston area of eastern North Carolina, CAP provided three auxiliary generators which furnished power for food freezers.

North Carolina's CAP mission headquarters was at Wilmington under the command of Lt. Col. Sydney Wilson, and for the Morehead area, Lt. Col. Dan Lilley was in charge. One hundred and fifty seniors and cadets participated, the officers reported, and there were five aircraft, 63 vehicles, 43 mobile radio units and 43 fixed radio units involved.

The eye of the hurricane passed off the Virginia Capes with winds in the Suffolk, Virginia, area estimated in excess of 100 miles per hour.

The entire city of 12,000 and the even more populous surrounding area were plunged into darkness. Hundreds of trees were uprooted, electric and telephone communications were disrupted, roads were washed out and whole communities isolated.

The National Guard and Suffolk's CAP Squadron were called on to help direct traffic, guard against live wires, clear streets of fallen trees and maintain radio communications.

The squadron's trailer-mounted 10-kilowatt generator was loaned to the Chesapeake and Potomac Telephone Co., to provide emergency power for the utility firm until employees could get their own emergency equipment in operation.

On up the coast, Connecticut loaned seven portable power generators to Civil Defense and itself had 200 members, 10 aircraft, numerous vehicles and mobile radio stations as well as six land rescue teams on the prowl. Rhode Island Wing, with 150 personnel at standby used 25 mobile radio units in a state-wide weather reporting and advisory service.

The specifics of how CAP is integrated with other civilian and governmental agencies aiding in disaster situations differs somewhat from state to state. Generally, however, CAP functions either as an agency of the state by virtue of an agreement between the governor and the CAP wing commander or as the result of state legislation giving CAP a "state hat" to wear or it goes into action wearing its "Air Force hat" as a major resource of the ARRS. In the case of the latter, local, area or state agencies request CAP assistance through the appropriate USAF rescue coordination center.

In California, for instance, a formal agreement between the wing commander and the State Office of Emergency Services (OES) provides the mechanism for requesting, authorizing and utilizing CAP throughout the state. The agreement provides that the resources of the wing may be employed either on a statewide or local basis with the wing commander retaining operational control at all times. The agreement also provides for reimbursement to CAP members for fuel and lubricants used on approved missions and insures that CAP members involved in providing support requested are covered under the provisions of State Workmen's Compensation. To facilitate the latter each CAP member is registered as a "civil defense worker". Included in the operating agreement are provisions that permit the OES to provide specialized training and to loan specialized equipment to CAP.

Among specific missions which the California OES sees for Civil Air Patrol include ground search and rescue, radio communications, aerial photographic and reconnaissance, light transport flights for emergency personnel and supplies, courier and messenger service, aerial surveillance of surface traffic and aerial radiological monitoring. In addition, CAP members and equipment can be employed in a wide variety of relief activities. Wing resources provide for flexible response to local and area requirements.

The largest of the nation's 52 wings, California has some 4,200 members—2,050 ground personnel, 950 observers and 1,200 pilots. The wing operates 48 corporate aircraft ranging from Piper PA-18 Supercubs to twin-engine Beechcraft C-45s. California CAP members provide the use of some 400 additional privately-owned airplanes. The wing has nearly 300 surface vehicles including 31 personnel carriers, 130 trucks, 18 busses and 80 four-wheel-drive jeeps. These which can move personnel together with supplies and equipment are augmented by more than 60 special purpose vehicles—25 heavy logistic rigs, two cranes, 14 ambulances, four fork lifts and 22 water buffalos (tankers).

It can deploy a 50-bed field hospital and 25 field kitchen units. For its own electric power requirements in the field or for loan to local agencies in emergency, the California Wing has nearly 250 gasoline-powered generators ranging from 18 huge, 10-kilowatt powerplants—eight of them fully mobile, the others portable—down to some 200 300-watt field units. Its statewide communications facilities include some 330 fixed stations operating on HF and VHF frequencies, 30 mobile stations and more than 100 aircraft stations (aircraft equipped to transmit and receive on Air Force and FAA aeronautical frequencies authorized for SAR use).

On a national scale, both the American Red Cross and the Salvation Army, this country's two major civilian disaster relief organizations, have established agreements with the Civil Air Patrol delineating methods and procedures for joint operations. In those emergency situations in which the Red Cross is active and where the operational control of CAP forces is under the appropriate USAF rescue coordination center, Red Cross requirements are communicated through the USAF mission coordinator and the matter of priorities where CAP forces are concerned is decided by the Air Force controller.

In those circumstances where the Red Cross must have emergency transportation by air of medical supplies, materials and Red Cross employees and where the emergency may not be of such a nature or magnitude to warrant Air Force participation, the agreement makes it possible for a Civil Air Patrol wing commander to respond, on a voluntary basis, to request for emergency services from a "duly authorized" representative of the Red Cross. In such a case, the CAP may then request from and receive reimbursement for out-of-pocket expenses for fuel, lubricants and commercial communications costs expended in performing the mission. Services which may be required by the Red Cross include, in addition to airlift, communications support in the form of both equipment and personnel ground transport and shelter manpower.

There is a flip-side to the CAP/ARC agreement which provides for Red Cross support to CAP forces under certain circumstances. In cases where CAP may be involved in search and rescue or disaster relief operations in remote or devastated areas, the CAP mission coordinator may request of the Red Cross support in feeding CAP forces, use of Red Cross disaster vehicles, assistance from Red Cross personnel in notifying next-of-kin where fatal crashes occur and in providing medical teams if their support is required at crash sites.

The most recent mutual-aid agreement—signed in late 1973—is with the Salvation Army and provides for approximately the same exchange of services as that with the Red Cross. The CAP will provide communications assistance, ground and air transportation and personnel to support relief efforts should the Salvation Army request it and, conversely, the Salvation Army will provide certain aid to Civil Air Patrol forces engaged in SAR or disaster relief activities in remote areas.

From earthquake-prone California and Alaska, through the mid-continent "Tornado belt", along the coast of the Gulf of Mexico and up the East Coast where hurricanes bring destruction on an almost yearly basis and out to Hawaii, target of the dread tsunamis (tidal wave)—the unpaid and often unheralded volunteers of the Civil Air Patrol are, on an almost daily basis, writing new chapters to what has become a living legend—a continuing story of heroism, self-sacrifice and dedication.

A CAP ground rescue team and its Air Force instructor show off their wares for community official during a disaster relief demonstration.

CHAPTER 11

People!—The "P" In CAP

"Red" Young of Tacoma, Wash., otherwise known as Lieutenant Colonel Young of the Civil Air Patrol's Washington Wing has "5,000 sons". That's the way Young, whose red crewcut now is mixed with grey, feels about it. You see, his only son died shortly after birth and for the past 20 years, Young has lavished a lot of care and affection on the Civil Air Patrol cadets who have been his responsibility. That's not to say that his six red-headed daughters and lovely wife, Pauline, didn't get their share for, from all reports, Red Young has a deep capacity for caring.

Now 53, Young served in the Marine Corps during World War II. He is a truck mechanic by trade and for many years raced motorcycles as a hobby. That, however, was before he learned to fly and discovered CAP cadets.

In 1958, the Washington Wing's emergency services training program needed a home. Red Young searched and searched, finally finding two World War II surplus barracks buildings nestled among tall trees near the end of the runway at Shelton's Sanderson Field. Some tall talking got CAP permission to use the buildings and through untold hours of back-breaking work on the part of Red, Pauline, other CAP senior members, cadets and their parents, chaos became order. Sleeping quarters, classrooms, a mess hall and kitchen took shape. Camp Young, as it is called today, became a going concern.

For 13 years, often at great personal sacrifice and expense, the Youngs, almost singlehandedly, kept Camp Young and the emergency services training program in gear. Red served as camp commander, instructor and director of the training program while Pauline served meals and cooked for the upwards of 100 hungry cadets and senior members who converged on the camp each month.

Earning a living and keeping the camp functional wasn't quite enough for Red Young. He saw a requirement to develop an organization which could provide direction and focus to the new capabilities resulting from the training being provided. He formed Washington Wing Team 3, a group of graduates representing various squadrons in the same geographic area.

Over a period of time, the team assembled the specialized equipment necessary to be self-sufficient in the field and to provide the mobility necessary to respond to a wide variety of contingencies. Radio direction finding equipment was procured and the team members trained in its use. The DF gear became an integral part of the team's mobile rescue vehicle. A truck also carries its own communications capabilities and a kitchen with its own water supply and equipped with a winch for heavy rescue work. A trailer carries portable winches, cables, special tools and additional rescue equipment while a second truck tows the "mountain goat". The team, under Young's direction, built the goat from an otherwise unusable jeep.

With Colonel Young at its helm, Washington Wing Team 3 has not only been involved in countless actual search, rescue and disaster relief missions, but also has travelled far and wide—to Hawaii, to the Air Force Academy at Colorado Springs and south to California to demonstrate its expertise. For the past 10 years the team—and invited guests—have spent Christmas vacation in a "winter survival mission" in the mountains of Montana deploying from a base camp at the University of Montana Forestry Camp near Missoula. Sometimes the group gets rather large, as many as 60 cadets along with participating senior members and cadets.

"The team is more than a rescue unit," Young observes, "this group learns about pride, confidence, teamwork, faith, fellowship and courage. They learn survival, canyon rescue, first aid and leadership. Their uniforms are impeccable and CAP history and tradition are as familiar to them as their birthplaces."

Just how much of Red Young's own personal philosophy and code of behavior has found its way into the hearts and minds of his young charges over the years is evident from this letter written to him by one of his cadets in 1968.

"Last December Bel-Tac during non-denominational services you asked us this Christmas not to run in under the tree and start ripping open the packages, but to first thank our parents for all they have done for us and to tell them how we loved them for it. Using your own case as an example, you said that when we became older we would realize and appreciate more fully all they had done for us, but maybe by then it would be too late to say all that we would have liked to.

"Sometimes you probably think that your efforts to convey your feelings and guide the cadets toward a better life and character are useless. Believe me, they are not. I took your advice last December and I thank you from the bottom of my heart that you gave it. This spring, I learned that one of my parents was seriously ill. Last night, this parent passed away. But because of your advice, I think I made this last Christmas a very happy and memorable one for both of my parents.

"So continue passing on life's teachings to us content in the fact that your efforts are worthwhile, even if they only help one person in a little way. Again, thank you very much for your guidance."

Since World War II, CAP membership has run between 60,000 and 80,000 with the average any one year about 70,000 senior members and cadets. In the late 1950s, the membership rose to nearly 80,000. Current membership (mid-1974) is some 60,000. Like golf, tennis, bridge, the National Guard, police reserve service or the local volunteer fire department, Civil Air Patrol membership isn't for everyone. Many adults and teenagers try it on for size and decide their primary interest is elsewhere. Many find that the unusually high level of commitment that is necessary is beyond their capability. Others find the regimentation and military-style discipline is not their cup of tea. Still others enter into CAP activities with such a fervor and zeal that in a few years they are "burned out", at least insofar as CAP is concerned.

However, for an uncommon number like Red Young, the Civil Air Patrol becomes almost a way of life. There are thousands of Americans today who have been working members of CAP for 20 years or more. There are hundreds of CAP families from Maine to Hawaii where the father, mother and youngsters are deeply involved in various aspects of the program. Two of these are the Burts of Northern Minnesota and the Framptons of Southern California.

Allan Burt; his wife, Lorraine; brother, Joe; brother-in-law, Elmo Crowe; and nephews, Bob Crowe and Ed Burt, all are senior members of the Grand Rapids Composite Squadron. Elmo has two sons, Jody and Farrell, in the CAP Cadet Program. Another son, Gary, was an active member until he passed away in 1969. Josephine Crowe was an active member until she was forced to terminate her CAP activities for personal reasons. Marvin, Ed Burt's younger brother, is a cadet.

The Civil Air Patrol found Allan, or perhaps it was the other way around, in the late 1940s. He became active in various squadron activities, including the cadet program, during the period when unit meetings were being held in a deserted CCC barracks which had been moved near the airport. As is often the case, this unit encountered a downhill streak which ultimately ended in its deactivation.

Undaunted by the demise of the unit, Allan's interest in CAP remained alive until, with the encouragement of officials of Minnesota Group 1 in nearby Duluth, he and Lawrence Randall reactivated the squadron. As could be expected, Elmo and Joe were among the first members of the reorganized unit.

The next step was formation of a flying club organized for the purpose of purchasing and operating a private aircraft— primarily devoted to CAP activities such as search and rescue missions and cadet orientation flights. In fact, it wasn't long before Allan and Joe were credited with a "save" in the club Cessna 150.

Since its reactivation, the Grand Rapids Composite Squadron has continued to grow and flourish. Joe Burt became its commander in 1971, a position he still holds. The positions of deputy commander for Cadets and Senior training officer behind him, Elmo now is deputy commander for Seniors. Allan is deputy commander for Cadets when he isn't busy earning enough money in refrigeration sales and repair to "afford his CAP habit".

Allan, Joe and Elmo are all CAP "mission pilots" while Rob and Ed have earned their FAA private pilot certificates and are working on their CAP mission pilot ratings. Six members of the Burt/Crowe clan hold CAP radio operator's permits and active CAP communications stations are in operation at both homes. Although Lorraine Burt's duty assignment is that of Squadron Information Officer—she has twice won the Minnesota Wing "Outstanding Squadron Information Officer" award—she also has emerged as the squadron's most active communicator, handling the bulk of the unit's traffic and meeting most of its Minnesota Wing net commitments.

Active Civil Air Patrol membership for Lt. Col. John "Kirby" Frampton of the California Wing; his wife, Dorothy; son, Dick; and daughter, Gail, totals a whopping 75 years. Dick Frampton, now a San Fernando Valley dentist, joined CAP as a cadet in 1950 and now, 24 years later, is also a Lieutenant Colonel and is assigned as Chief, Plans and Programs, California Wing.

His father, Kirby, was until very recently director of Senior Training on wing staff. With nearly three decades of active service, Kirby talks glibly about "retiring from CAP." Close friends merely smile and say with knowing grins, "sure you will!" Dorothy Frampton, a veteran of some 22 years as a CAP member, still is active. Gail, however, left the program after completing five years as a cadet and upon taking on the responsibility of a husband and family. In addition, Kirby's brother, Robert, was an active member for a large number of years.

Kirby Frampton first became a member of CAP a little more than a year after it was founded in 1941, and has been active for 27 of the intervening 32 years. His World War II affiliation was with the Boulder City, Nev., squadron. Later, in California, Kirby and Jerry O'Brien activated the Rosemead Composite Squadron in the San Gabriel Valley, east of Los Angeles. Subsequently, the pair went on to form California San Gabriel Valley Group 15.

Advancing to the wing staff, Kirby served as Southern Area Director of Cadets, Wing Director of Cadets, and Wing Director of Senior Training while son, Dick, moved up through the cadet ranks. Young Frampton was instrumental in the establishment of the California Wing Drill Competition program. Sister, Gail, went on to command the winning cadet drill team in several competitions. As a cadet and senior member, Dick has attended 13 summer encampments on Air Force bases, while his father has attended 10 as a senior staff member, three of these as encampment commander. Taking time out to attend and graduate from Dental School, Dick entered the senior program in 1960 first with Group 15, his father's old unit, then moving to the California Wing Staff serving as Assistant Director of Cadets, Director of Cadets, Chief of Staff, Wing Inspector and Chief, Plans and Programs.

Born in Kansas, graduated from the University of Alaska, additional schooling in Hawaii and Peking, China, visited all 50 states living in five of them, traveled in Europe, North America, the Orient! You might say that Jamie Cameron, now of Kaneohe, Hawaii, has been around a bit, seen a few things and experienced a few others. Oh, yes, she also was in Hawaii when the second World War became a hot war for the U.S. and for nearly five years was a volunteer "night driver" for the Red Cross Motor Corps in Honolulu.

A Civil Air Patrol member for 17 years, Jamie, a teacher and now Deputy Director of Materiel for the Hawaii Wing, says:

"I've met a lot of people in my travels, but none have added more interest and pleasure to my life than those I have met through my membership in the Civil Air Patrol. The only experience I have had that has given me as much satisfaction as my CAP activities was driving for the Red Cross during World War II."

Lt. Col. Jamie Cameron's contributions to the Civil Air Patrol program also have been considerable proving that CAP membership really is a two-way street. Jamie's professor, the famed scientist, Dr. Edward Teller, and CAP's aviation education program all had a part in bringing her into the fold. In 1957, CAP scheduled a national aviation education workshop in cooperation with the University of Hawaii, one of many over the years designed to equip teachers to better meet the educational challenges of the aerospace age. Jamie, being a teacher, was interested, but not as much in the aviation portion of the program as in hearing and getting to meet Doctor Teller. Not one to do things half way, Jamie not only enrolled in the workshop—achieved her desire to meet and hear Teller—but also became a CAP member.

"I was assigned as a senior member," she recalls, "to teach 'the books', as the cadets called them, with the Drum and Bugle Corps Squadron. As you may have guessed, I neither drummed nor bugled and was the only female in the squadron. It was two years before a flight of girls finally was added."

Since 1957, Jamie has attended a second national workshop—this one at the University of Montana where she took up flying as her extra activity. She graduated from the Civil Air Patrol Staff College, Maxwell AFB, Ala.; attended a CAP National Convention in Washington, D.C., and three regional conventions; traveled to Europe with a University of Tennessee CAP group; and has been on the staff of 10 Hawaii Wing cadet summer encampments and three cadet flying encampments. In between special programs, Jamie Cameron wrestles with the wing's supply problems and, because she still apparently does not have enough irons in the fire, has taken it upon herself to see that every one attending wing staff meetings is fortified with a steaming cup of coffee, cream and sugar and occasional donuts. Rather, the proceeds from her cup go into a scholarship fund for the wing's flying encampments.

Speaking of contributions to achievement of the Civil Air Patrol's objectives and of the distaff side automatically brings one to Clara Elizabeth Livingston. For nearly two decades, Puerto Rico has represented one of CAP's most active and most effective wings. The largest share of the credit goes, of necessity, to CAP Colonel Livingston. Now actively involved in Florida real estate where she still is an active CAP member, Clara Livingston called Puerto Rico home for most of her life and it was above this island home she first got bitten by the aviation bug.

In 1927, the same year that Charles Lindbergh became a legend in his own time flying the Atlantic, young Clara flew over the Puerto Rican capital, San Juan. In those brief moments, she became enamoured of flight and began a love affair with the airplane that still continues. Within three years of that flight, she graduated from the Curtiss Wright School of Aviation—then headed by the legendary Casey Jones—returning to Puerto Rico to construct a landing strip on the family plantation at Dorado. She soon became the proud owner of a Rearwin Ken Royce, open-cockpit biplane and then there was no keeping her on the ground. In the summer of '32, for instance, she and a friend Marian Wolf, climbed aboard the Rearwin and took off for San Diego, Seattle, New York and returned to San Juan. This nation-spanning, round-robin was only one of scores of long-distance, over-water flights Clara and friends packed into the 30s. During this same period, she started a flying school at Dorado and became a friend and confidant of many of the greats of early American aviation, among them Amelia Earhart. Amelia and her navigator, Fred Noonan, rested at the Livingston home on the first leg of their ill-fated world-circling flight in 1937.

Clara Livingston's application for CAP membership is dated December 1941 which, some three-plus decades later, would qualify her as a charter member. At the time, she was completing a flight instructor course in the states. Despite a heavy schedule as an instructor during the war, she still found time to fly as a search and courier pilot with CAP. Two years in the Women's Army Corps followed three hectic years as a flight instructor after which she returned to Puerto Rico to operate a flying school sandwiching in the time necessary to obtain a multi-engine rating for fixed wing aircraft and commercial helicopter pilot license.

May 24, 1956, Colonel Livingston became commander of the Puerto Rico Wing, a position she held through 1972. One of her first priorities was to begin a program to see that aviation education became part of curriculum in the Puerto Rico public school system. This was a challenge and a half.

Notwithstanding the fact that the economy—in fact perhaps the destiny—of the United States has been inexorably affected by aviation since the brothers Wright flew over the sands of Kitty Hawk in 1903, persuading educators to provide aviation-oriented subjects has been an up-hill battle. Introducing aviation education into the schools of the Commonwealth of Puerto Rico was no exception. Then again, perhaps it was an exception, for Clara Livingston met with outstanding success. Achievement did not come without hard work, but when it came it was complete. Aviation education—based on the CAP-developed program—became a formal elective in the curriculum of all the Puerto Rican public high schools and by the mid-1970s more than 60,000 young men and women had been exposed to the impact of aviation on their lives, the future of their commonwealth and the future of their nation.

In retrospect, Clara Livingston believes that the Civil Air Patrol has "greatly benefited" the people of Puerto Rico "by bringing together as a team, citizens of the island and of the continent."

"Islanders," she says, "naturally tend to be insular and somewhat parochial in their attitude toward the outside world. CAP has played an important role in erasing this gap with its programs—for young and old alike—which have helped build a bridge of understanding. The first and most important benefit derived from CAP's programs—is a more firm comprehension of national unity."

"This is particularly significant for our cadets," she adds, "especially those who have been privileged to visit the United States and foreign countries. And all cadets are exposed, through their CAP studies and activities, to experiences which broaden their horizons. The entire membership in Puerto Rico has been given a front row seat and an opportunity to participate in the growth of aviation. This is essential to all island citizens because it was the air age that ended Puerto Rico's isolation and has now linked to all parts of the world with bustling air traffic."

Throughout the Civil Air Patrol, the aviation education program in Puerto Rico is considered a "showcase" for the CAP. In the words of Brig. Gen. Richard N. Ellis, USAF, a recent past CAP National Commander: "Clara Livingston has been a driving force and pivotal figure" in creating such a showcase. Holder of virtually every medal and honor of any consequence that CAP can bestow, Col. Clara E. Livingston was, in 1972, voted into and installed in the Civil Air Patrol Hall of Honor. This singular recognition has been added to accomplishments like being the first woman pilot of Puerto Rico, first woman multi-engine pilot of Puerto Rico, first woman helicopter pilot of Puerto Rico and first woman to direct a government-approved flying school in Puerto Rico.

Clara Livingston is not the only woman to become a Civil Air Patrol wing commander. In CAP's more than three decades there have been four—Livingston in Puerto Rico, Col. Nanette Spears in New Jersey, Lt. Col. Mary C. Harris in Idaho, and Col. Louisa S. Morse of the Delaware Wing. Colonel Spears lost an untimely battle with a fatal illness in the late 1960s, but for Louisa Morse it still is business as usual in the Delaware Wing headshed. She has commanded the wing since 1953 and has been an active CAP member since early in World War II.

Joining CAP in 1942 as a private, Louisa Spruance Morse rose through the enlisted and officer ranks reaching the point where she is privileged to wear the silver eagles reserved for colonels. She admits she has performed perhaps every job in CAP beginning as an instructor—she won her aviation ground school instructor's rating just prior to becoming a member.

Command has always come naturally to Louisa Morse. She is the daughter of a distinguished World War I Army Officer, Col. William W. Spruance, Jr., and her brother, Brig. Gen. William W. Spruance, was adjutant general of the State of Delaware. Around the house, Louisa outranks her husband, but not by much. He is Lt. Col. Albert W. Morse, Jr., CAP, formerly Lieutenant Colonel Morse of the U.S. Army.

The Delaware Wing commander runs a tight no-nonsense organization and its record in competition with the other 51 wings of Civil Air Patrol consistently shows it. For instance, when it became popular for young men to assume long hair styles and conform by becoming dedicated non-conformists, Colonel Morse remained firm in her resolve that CAP was "a military-oriented organization and as such its members should look like it." When male hair began to get too long, she issued a simple, straightforward order—

"Get your hair cut or don't fly!" Since the Delaware Wing is a flying organization from top to bottom, that order had the desired effect.

Louisa Morse's philosophy is as straightforward as her command posture. "In our program, she explains, we work with fine, intelligent, young cadets—boys and girls who are eager to learn and to contribute something worthwhile.

"So much publicity is given teenage hippies, long-hairs and protesters," she says, "and the 'nice' kids are overlooked. But not in the Delaware Wing. We have nearly 300 cadets in this state. They enjoy the military drill and classroom instruction as much as they do the opportunity to fly as passengers or pilots. They are well groomed, neatly uniformed young people. Courtesy and military discipline become automatic with them. They are living proof that most young men and women today welcome and respond to firm intelligent supervision and leadership."

One of the Civil Air Patrol's most colorful members holds several national titles and responsibilities—National FAA Coordinator, National Safety Advisor and National Chief Check Pilot. From this information you might deduct that Col. Edwin Lyons of West Hempstead, Long Island, N.Y., is somewhat familiar with flying machines and aeronautics. Well, he is. With some 32,000 pilot hours to his credit, his aeronautical certifications look like this:

FAA Airline Transport Pilot (airplane, single and multi-engine, land and sea; rotorcraft, helicopter, glider)

FAA Certified Flight Instructor (airplanes, gliders and instruments)

FAA Commercial Pilot Examiner and Instrument Pilot Flight Examiner

FAA Parachute Rigger

FAA Airframe and Power Plant Mechanic

FAA Ground School Instructor (advanced)

FAA Free Balloon Pilot

Republic of China Commercial Pilot (airplane, single and multi-engine, land and sea)

Ed Lyons, who was inducted into CAP's Hall of Fame in 1973, learned to fly in 1928 and free-lanced as a pilot through 1935 when he began operation of a flying school at historic Floyd Bennett Field on

Long Island. Itchy feet and the quest for adventure took him to Spain in 1938 where he flew in combat as a contract volunteer for the government forces. From Spain, Lyons went on to Israel (then Palestine) where he owned and operated the first flying school and flight service in Tel-Aviv.

Back home in the states in 1941, Ed became chief pilot for Lincoln Aeronautical Institute, Lincoln, Neb., and two years later accepted a position as Director of Flight Training with the Mountain State Aviation Corp., Denver and Boulder, Colo. He entered government service in 1944 with the Civil Aeronautics Administration and was named a Senior District Flight Training Supervisor. However, Ed lyons still wanted to be his own man, so in 1946 he took the big plunge and organized Lyons Flying School at Zahns Airport, Amityville, Long Island. Hard work and long hours both in the air and in the office paid off and in 1950 when operations at Zahns Airport were consolidated, Lyons became co-owner of the airport and president of Amityville Flying Service, Inc., with duties as airport manager and company administrator. In the following two decades, this school and airport operation has grown into one of the largest in the eastern United States providing flight instruction under contract to both the state and Federal governments.

Like Clara Livingston and Louisa Morse, Ed Lyons is virtually a charter member of CAP. He has seen duty at flight, squadron, group, wing and regional level—past commander of the Northeast Region—and is a Civil Air Patrol command pilot.

Civil Air Patrol members represent virtually all races, creeds, ethnic backgrounds, religious and economic backgrounds. It has been truly integrated from its earliest beginnings and thus accurately can be described as a microcosm of American life. It is quite evident that it offered the opportunity for women to work for and achieve whatever level of personal accomplishment and prestige their individual desires and capabilities would permit long before the days of "women's lib". It also has been generally understood that, by and large, the senior command echelon—wing and regional commanders—generally were from an economic class which permitted them to devote large amounts of time and spend large sums of money in performing their duties. Undoubtedly this was true to a great extent in CAP's early days when it was possible to draw an accurate profile of a wing commander by including such benchmarks as high level professional men or independent businessmen, socially prominent; in a relatively high income bracket. Not so today. Take this random sampling of top commanders for instance:

Mississippi—Col. John A. Vozzo - research physiologist

Tennessee—Col. William C. Tallent - elected official (politician)

West Virginia—Col. Robert E. Gobel - industrial planner (FMC Corp.)

Minnesota—Col. John T. Johnson - working commercial pilot

Wyoming—Col. Albert Lamb - automobile salesman (left law enforcement because the hours "conflicted with my CAP work")

Virginia—Col. Randolph C. Ritter - long-haul truck driver

Arizona—Col. G. Eugene Isaak - lawyer

Arkansas—Col. Bob E. James - businessman (owns and operates three taxi cab companies)

North Carolina—Col. Ivey M. Cook, Jr. - business executive (president of Cook Body Co.)

New York—Col. Paul C. Halstead - businessman (operates an aircraft storage complex at Islip McArthur Airport)

Utah—Col. Larry D. Miller - salesman (HI-Land Dairy)

Texas—Col. Joseph L. Cromer - insurance agent

Puerto Rico—Col. Rudolfo D. Criscuolo - businessman

Massachusetts—Col. Carl J. Platter - administrator (executive vice president and general manager, Alcomm Service, Inc.)

North Central Region—Col. William H. Ramsey - business executive (president of Wilson Learning Corp.)

Rocky Mountain Region—Col. Frank L. Swaim - airline captain (flying jumbo jets)

North East Region—Col. Julius Goldman - business executive (president Revere Aviation, Inc.)

It is interesting also to note that the chairman of the Civil Air Patrol National Board, Brig. Gen. William M. "Pat" Patterson, CAP, is a sales representative in the heavy equipment line. In other words, he works for a living.

Like the rank and file of CAP, its leadership also is drawn from all walks of life. And like the rank and file of CAP, the men and women who lead it have several things in common—an abiding faith in the American way of life, a strong feeling of moral responsibility toward the young people of the country, a conviction that the United States must maintain supremacy in the air and in space if it is to remain a nation of free citizens and a willingness to make what ever personal sacrifices are required—even the supreme sacrifice—in the preservation of human life.

As one is exposed to the history as well as the current exploits of this Air Force civilian auxiliary, it becomes increasingly apparent that there are an unusually large number of women involved—involved in all phases of CAP activity from comparatively passive administrative duties to its high-risk operational missions. As it has been pointed out elsewhere, one of these areas is communications where women have become the backbone of the nation-wide network. Another is disaster relief, particularly that phase which calls for first aid and nurses training. Three women come to mind as outstanding representatives of this group. They are Minnesota's Lt. Col. Laura M. Black, Puerto Rico's Maj. Irma C. Irizarry and California's Lt. Col. Jane Hedges.

A roly-poly ball of energy who admits to "taking nurses training sometime in the 30s", Jane Hedges is known far and wide as a "fireball" and the pace she sets bears this out. Eighteen years in Civil Air Patrol, 32 years in the American Red Cross and Civil Defense,

Jane currently is a special assistant to the Pacific Region commander for disaster relief. Concurrently, she is a working member of the Disaster Relief Committee of the Los Angeles County Medical Association and thus is the prime mover in bringing together CAP forces and medical organizations in one of the nation's most populated counties.

Jane took her nurses training at Peter Bent Brigham Hospital, Boston, Mass., and not too long thereafter moved to California. Additional specialized courses at Fresno Junior College—now Fresno State—followed and this developed into a teaching assignment in the college's adult education department. Once Jane got a taste of teaching the appeal of regular nursing paled. More specialized instruction, more specialization in the scope of her teaching assignment followed.

Now a move to Los Angeles where she began a long association with the State as an instructor in radiological monitoring, medical self-help and personal and family survival. More and more she moved away from nursing skills as they are practiced in the hospital specializing in applying those skills to the field emergency environment with the result that most of her adult life has been occupied with constantly acquiring new specialized nursing skills—keeping up with the state-of-the-art in emergency techniques—and then passing those skills on to others. More and more in recent times, Civil Air Patrol senior members and cadets have been her students. In 1974, Colonel Hedges was honored on national television—ABC's "Girl In My Life" program—for her contributions to CAP in this area. Featured on the program were cadets who had saved the lives of others—one his own father—with the skills learned in Jane Hedges' first aid training classes.

Like Jane Hedges, the North Central Region's Laura Black continues her long years of giving of herself and her skills to CAP—it is now more than three decades since she first became a member. As is so often the case with those unusual individuals who do more giving than taking, Laura has found time to make outstanding contributions in many areas. After graduation from the University of Minnesota where she received her RN, Laura Black went on to the Harvard Medical School for advanced training in physical therapy. Back in Minnesota, she put that training to work and is widely known for extensive work in this field. She is director of Physical Therapy at the Lindsey School for Crippled Children; a physical therapist at Gillete State Hospital, and is active in the Visiting Nurses Association. Not satisfied with just bringing to CAP her nursing expertise, Laura Black also added a private pilot's license and an FCC amateur radio operator's license.

"Mimi" Irizarry—that's what they call this Puerto Rican-born mother of three—has in a relatively short time become what her associates in CAP call a "fixture" in the Puerto Rico Wing. They hasten to explain that by fixture they don't mean that Mimi Irizarry is either unnoticed or taken for granted. Rather, she is very much in evidence "patching up and ministering to the cuts, bruises and ills of all those who participate in wing civil defense and search training missions as well as the real thing". She has instituted a "preventative medicine" program and for members who elect to participate she regularly samples and keeps records on blood pressure, pulse rates, temperatures and other physical parameters.

Formerly a surgical nursing instructor at the University Hospital School of Nursing, Mimi currently is Interagency Coordinator, Summer Child and Youth Activities in the Commonwealth Office of the Secretary - Social Services. She is an instructor in medical self-help not only to CAP but also to the Army and Air Force ROTC, the Antilles Military Academy, the American Military Academy, the Puerto Rico Police Special Reserve Force and the Home Economy Clubs. A specialist in civil defense, she is a graduate of the Office of Civil Defense Staff College and has completed advanced courses in CD Management, CD Plans and Operations, Disaster Management, Shelter Management, and Disaster Nursing. She is a first aid instructor for the American Red Cross, the Boy Scouts and the Girl Scouts. Still another side of Mimi's character is evident from her active participation in organizing field trips for mentally retarded children and her deep involvement in the Department of Education's child school health program.

At any given point in CAP's 30-plus-year history its rolls have reflected the names of many of the nation's foremost women pilots as well as hundreds of others whose motto is "I'd rather be flying!" There is a good reason for this. CAP, from the outset, offered women the oppoetunity to do a man's job in a world where by-and-large the female sex was relegated to the kitchen or the nursery. This was particularly attractive to those women who developed a taste for aviation in the 1920s; those who became pilots during World War II serving either as flight instructors at military contract schools or as members of the WASP (Women's Air Force Service Pilots), part of the Ferry Command, delivering everything from fighters to bombers from the factories to operational bases and overseas staging areas; and those since who have become pilots.

Among them have been Jacqueline Cochrane, Louise Thaden, Jean Ross Howard, Clara Livingston, Jean Adams Cook, Rosemary J. Lane, Selma Cronan, Lola Perkins, Ada R. Mitchell, Dorothy Tuller, Lucille Cantway, Virginia Jansen, Clair Griswold, Nola Henderson, Evelyn Bryan, Edna Whyte, Georgiana McConnell, Coral Bloom, Jerri Cobb, Pat Davis, Ruth O'Buck, Nancy Howard, Beverly Harp, and scores of others.

And then—there was Pearl White Ward of Lake Charles, La. Now Pearl (White until she met and married Excell Ward) was more than just one of the nation's pioneer women aviators—she learned to fly in 1930—she also was one of aviation's pioneer professional parachutists and for several years led an adventurous and monetarily rewarding life as an exhibition jumper in the golden days of the barnstormers.

The chain of events which led to Pearl becoming a pilot, parachutist and ultimately an active CAP squadron commander, began when she left her home in Smithville, W. Va. Her family sent her to Washington, D.C. to attend secretarial school. For the first few weeks 16-year-old Pearl attended school and in general led the kind of life her family had in mind. But one day, she read an ad in a Washington newspaper. It said: "Learn to fly for $45."

The next day Pearl didn't go to school. Instead, she headed for Beacon Airport in Alexandria, Va. There, she said she was 21 (then the minimum age for taking flying lessons) and paid her $45.

Pearl caught on fast and soloed after only two and a half hour' flying time. She never went back to secretarial school. Instead, she spent a great deal of her time around Beacon Field.

One day one of the pilots asked her if she'd like to make a parachute jump the following weekend. Pearl thought it sounded exciting and said "Yes." For days before the Sunday she was to jump, Pearl was heralded by the local newspapers as "the first bird-woman."

Of course, she was scared, but she was excited, too, and glad to have the opportunity to learn something new.

All she was told was "Jump. Count to 10 and pull the ripcord." When she jumped, she pulled the ripcord right away and forgot all about counting. Her landing wasn't the most graceful in the world, but she wasn't hurt and the jump was a big success.

On landing, another pilot, "Wild Bill" Haney, came up to Pearl and gave her two parachutes and some advice.

"Pearl", he said, "If you want to continue jumping, take these two chutes. I know they're OK and you won't have any trouble. Don't ever jump unless you have one of these two chutes on and unless you know it's been packed and inspected by government inspectors."

That first jump started Pearl on a new, exciting career. She followed the air shows and did stunt flying and parachute jumps (for several hundred dollars a jump). Those air show days were fun—full of money and publicity and excitement. But Pearl never forgot the advice of Wild Bill Haney; she never jumped unless she used one of the chutes he'd given her and unless she knew it had been properly packed and inspected.

It all changed on Sunday, Aug. 24, 1934.

That day, Pearl was scheduled to jump in an air show at Galax, Va. She hadn't had time, between bookings, to get her two "lucky" chutes packed. She had made up her mind not to jump that day. Rain canceled most of the show, but when the weather cleared about 4 p.m., her manager came and asked her to make one jump for some of the spectators still on the field—to arouse their curiosity and make them want to return to see the full show the following Sunday.

Pearl explained that she hadn't had time to get her chutes packed. The manager told her she could use one of his chutes. Remembering Haney's advice Pearl said "No." Then the manager began testing her saying she was using the chute story as "an excuse", that she was "afraid to jump."

Pearl saw red. She put on the manager's chute, climbed into the plane with him. In the air, premonition gripped her. She felt sure the chute wouldn't open, that she would be "splashed all over the ground" in front of her thrill-seeking audience.

She had "just about decided to jump" when suddenly she was out of the plane. Pearl still can't say whether she was pushed or whether she jumped. She pulled the ripcord. The chute wouldn't open. She pulled again. It still didn't open. Pearl saw the ambulance racing across the field. She heard the spectators scream.

Once more she pulled the ripcord. The chute opened just before she hit the ground. She remembers the impact, bouncing up in the air and then being caught in the arms of a male spectator.

A week later, Pearl awakened in a hospital near Galax.

"It was such a pretty room, with nice curtains and flowers, that I was sure that I had died and gone to heaven," Pearl says.

She suffered severe back injuries and was in a cast for several months. She never made another parachute jump, although she did start flying again a year later.

Not all Civil Air Patrol members are former servicemen or women, pilots, licensed communicators, nurses or doctors. Hundreds, perhaps even the majority do not even have aviation backgrounds. They simply are citizens to whom the basic goal and objectives of the CAP appeal. Many of them began their Civil Air Patrol careers as cadets. In this manner, the organization is, in fact, self-renewing, self-vitalizing.

A case in point is California's Maj. Barbara Ferguson, commander of Group 1 in the San Fernando Valley. Still only 31, Barbara joined CAP as a youngster in 1956. Where some women remember their teen years in terms of who they dated or what famous movie star they had a crush on, Barbara remembers the cadet summer encampments and the two years she was a member of the Pacific Region championship drill team. For her, it was natural to move on into the senior program when she reached maturity. Her first assignment as a senior was to command the squadron in which she began as a cadet. Assignments at group and wing level followed but she felt too far away from the action and elected to take over a struggling all-girl cadet unit, Squadron 63. In three years, she built the squadron from nine cadets to more than 30, a unit twice named the Outstanding Cadet Squadron in California, much to the chagrin of the wing's male cadets.

Not only has Barbara Ferguson distinguished herself in CAP, she has established an even more enviable track record in civilian life. Born in Memphis, Tenn., she got her bachelor's degree in Business at Pepperdine University, Los Angeles, and went on to obtain a master's degree in Business Administration from the same institution. When she is not performing her duties as executive for Bomar TIC, Inc.,—Materials Control Manager for plants in Newbury Park, California and Helena, Ark., with overall responsibility for purchasing, production control, material control and quality control in this high-technology company—or with her duties as a CAP group commander, she teaches business and industrial management at Moorpark Junior College.

Only five feet five inches tall, weighing 128 pounds, the major doesn't look her age, particularly in field uniform surrounded by a staff made up mostly of six-foot males. She does, as one lieutenant put it, look like a "slip of a girl". Developing the theme, another adds:

"Have you noticed how those blue eyes take on a steely, grey glitter and how that small chin goes hard as a rock when she chews you out? When Barbara Ferguson chews you out, believe me, you know you've been chewed."

One thing her staff and the members of the squadrons who regularly stand the major's inspection agree on, "if Major Ferguson chews you out you can be sure you deserved it."

Her sector commander, Lt. Col. James Barnes, points out that his Group 1 commander was named the California Wing's 1973 Outstanding Senior Member, adding that "she is one of the finest CAP commanders I have ever had the pleasure of working with. She doesn't ask or expect her people to do anything she can't or won't do. She's a leader all the way and people will follow her anywhere."

In the final analysis, CAP is all the Barbara Fergusons, Red Youngs, Allan Burts, Kirby Framptons, Jamie Camerons, Clara Livingstons, Ed Lyons, Frank Swaims, Randy Ritters and Pat Pattersons. CAP is people and it is people that make CAP. All CAP people are not famous. Nor are they all individually outstanding. They all are not necessarily heroic. They all are not adventurous. Service in Civil Air Patrol does, however, take a certain special willingness to give that extra few hours of time when it appears to have run out, that extra bit of effort when there seems to be none there and those few extra dollars for which there were many other places to go.

This isn't to say the people of the Civil Air Patrol are not rewarded for their service. They are. Their reward is satisfaction—that satisfaction that comes only with serving your fellow man by doing something you really want to do.

CHAPTER 12

The Eagle's Nest

"I dare you to have your hair cut and not wilt under the comments of your so-called friends. I dare you to clean up your language. I dare you to honor your mother and father. I dare you to go to church without having to be compelled to go by your parents. I dare you to unselfishly help someone less fortunate than yourself and enjoy the wonderful feeling that goes with it.

"I dare you to become physically fit. I dare you to read a book that is not required in school. I dare you to look up at the stars, not down at the mud, and set your sights on one of them, that, up to now, you thought was unattainable. There is plenty of room at the top, but no room for anyone to sit down. Who knows? You may be surprised at what you can achieve with sincere effort. So get up, pick the cinders out of your wounds, and take one more step.

"I dare you!"

These words were written in 1964 by a former cadet member of the Grand Forks, Neb., Civil Air Patrol squadron. He was at that time still little more than a youngster. The words are part of a letter written to the Grand Forks Herald entitled "Challenge to Youth." They were penned by Clifton Cushman in the period immediately after he suffered perhaps the most crushing disappointment of his young life.

Very early in his youth, Cliff Cushman set his sights on the Olympic Gold Medal. His tenure in the body-and-mind-building CAP cadet program was a step toward this goal. In 1960 he came within a hand's grasp of that goal winning a silver medal in the 400-meter hurdles at Rome. In 1964, striving for a second chance—this in the Los Angeles trials for the U.S. Olympic Team—he stumbled and fell. Cliff hoped to try again but he never got the opportunity. That inexplicable confrontation called Viet Nam interfered. On September 25, 1966, Air Force Capt. Clifton Cushman was piloting his jet beyond the Demilitarized Zone. He was shot down over North Viet Nam. Officially, he is missing in action.

Cliff Cushman delivered his challenge to Young America circa 1960s. Well over a decade later it is perhaps even more pertinent than at the time it was issued. Cushman certainly could not have foreseen the further changes in philosophy, conduct, ethics and basic morality that were to create an ever-deepening chasm between a large segment of American youth and those principles held dear by the founding fathers.

Civil Air Patrol cadets are not the only young Americans today who still ally themselves with our traditional ethic—family, God and country. They are, however, in the vanguard and they number in the tens of thousands. Over its 30-plus year history, cadet membership in CAP has averaged from 25,000 to 35,000 active members in any one year. Basically, a five-year program, an estimated 5,000 or more "graduate" annually taking their place in adult society—many of them continuing CAP affiliation as senior members while pursuing their individual career objectives. There are few records of cadet membership from the World War II years, but based on those available since 1948, there are more than 130,000 who now are taking their places in industry, business, government and the professions. These men and women certainly are among those who, like Cliff Cushman, have remained true to traditional American principles.

One of those is Frank Borman, Vice President, Operations, Eastern Air Lines. Most Americans recognize Borman for another contribution to society—they remember Air Force Col. Frank Borman, astronaut, command pilot of Gemini 7, the 14-day "endurance" space flight that set the stage for later voyages of Sky Lab, and as the command module pilot on Apollo 8, the mission that paved the way for Neil Armstrong's historic lunar landing.

Another is Maj. Konrad Trautman, USAF, who, like Cushman, was shot down over North Viet Nam. Unlike Cushman, Trautman's whereabouts are known today. He is on active duty with the Air Force, electing to remain in that status after spending five and a half years as a prisoner of war. One of Trautman's pleasures on returning to a grateful homeland was to find that in the midst of a "changed world" his son, Konrad, Jr., and daughter, Diane, "had grown up to what he remembered". It could also be that the fact both are active CAP cadets in Steelton, Pa., and during his imprisonment proudly wore the Air Force blues has something to do with Trautman's sense of pride and well-being.

Still another CAP Cadet alumnus "making good" is Airman First Class Diane Scobee, former cadet commander of Manhattan Squadron 4, New York Wing, CAP. That airman first class title is not one bestowed by CAP, but rather by the USAF since Diane now is one of the few women helicopter mechanics in the Air Force. Most recently with the 55th Rescue and Recovery Squadron, Elgin AFB, Fla., Diane has earned the praise of her supervisor who says she is performing as well as the other airmen. Proponents of women's lib might take issue with the choice of words and manner in which this bit of praise is given, but if it is examined in the context of a tough Air Force maintenance supervisor on whose shoulders the safety of scores of others depends it emerges as significant indeed.

Former cadet, Scobee, demonstrated the same quiet tenacity on entering Air Force basic training at Lackland AFB, Tex., she exhibited during her rise to cadet commander of her CAP squadron at home. On arriving at Lackland, she "stressed" to military evaluators the fact she was "determined to work on an Air Force flightline." It wasn't easy, she recalls, and admits encountering "some difficulty" convincing her superiors she was "serious" about becoming a mechanic.

Graduating from technical school, Diane found herself the center of no little attention once she was on the line. Accepted at first with some reservations by her co-workers, she soon proved she could hold her own in this predominantly male career field. Some of her success is due to the level-headed attitude she expresses this way:

"I believe I can accomplish most of the jobs on the flightline but I am not going to try and do something that is over my head just to prove a point."

Two other former CAP cadets who are making good use of technical and leadership experience they gained in the program are Bob Hickox of Rexford, N.Y., and Mike Snedeker of Chevy Chase, Md. Snedeker is a second lieutenant jet pilot trainee in the Air Force having graduated with the class of 1974 from the U.S. Air Force Academy. Hickox, still a cadet second class at the Academy, will graduate in 1975 and enter active duty as an Air Force engineering management specialist.

Snedecker and Hickox are among the 750-odd former CAP cadets to have graduated from the Academy or who are currently enrolled there. All things being equal, approximately 11,000 young men will have graduated by 1977. When the stringent mental and physical requirements for entrance to the Air Force Academy and the high standards that must be maintained are taken into consideration, the fact that, since 1959 when the first class was graduated, nearly seven percent of the graduates have been former Civil Air Patrol cadets assumes considerable importance. Through 1973, there were 509 former CAP cadets among the academy graduates. Including the Class of 1977, another 242 are slated to complete their studies and become officers in the USAF.

Brig. Gen. Hoyt S. Vandenberg, Jr. (son of the former Air Force Chief of Staff), who currently is Commandant of Cadets at the academy, puts it in perspective:

"The mission of the United States Air Force Academy is 'To provide instruction and experience to each cadet so that he graduates with the knowledge and character essential to leadership and the motivation to become a career officer in the United States Air Force.' We provide cadets with the military, academic and physical education necessary for them to graduate with the character and leadership potential so necessary in this aerospace age.

"The Civil Air Patrol cadet program is an outstanding 'breeding ground' for young men who desire acceptance into the United States Air Force Academy. It teaches and represents all the attributes in young men that make for a successful career at the academy and in the Air Force." According to Vandenberg, "CAP cadets learn about a military chain of command as they rise through noncommissioned officer and officer positions. They have the opportunity to participate in leadership activities, opportunities where young men help plan squadron projects then lead their peers in the accomplishment of assignments. Finally, CAP cadets are in close proximity to an aviation environment learning about military aircraft as well as participating in operational flying activities such as search and rescue training and the earning of FAA pilot licenses. Former CAP cadets are noted for their performance in cadet wing positions at wing, group and squadron level here at the academy."

"Our observation", Vandenberg adds, "is that CAP cadets learn quickly, apply themselves readily and demonstrate good physical condition. It is obvious to me that the Civil Air Patrol is one of the most beneficial programs available to young men in America. It teaches leadership, maturity, self-confidence and self-discipline."

The general strongly recommends the Civil Air Patrol cadet program to anyone considering a nomination to the Air Force Academy.

But what of the cadets themselves. In retrospect, what do they think about their experience in CAP and its bearing on current performance. Bob Hickox credits his cadetship in CAP with steering him "toward the most important step" in his life.

"If I had not been in CAP, I probably would not have sought an appointment. Through support from my squadron commander and my parents, I was able to secure an appointment to the Class of 1975. Since entering the academy I have appreciated the quality of CAP training more and more."

"Civil Air Patrol", Hickox explains, "has contributed immeasurably to my physical and mental growth. The most important lesson I learned in CAP was self-discipline. I was compelled to budget my time, to place important tasks first, to accomplish the chores I had to do. I learned many other things as well. The aerospace education I received has been invaluable in my private pilot training and in my work here at the academy. Military training I received as a CAP cadet taught me a sense of disciplined duty; it was a great help when I arrived here.

"Another plus I received from Civil Air Patrol was my private pilot's license. CAP motivated me toward achieving that goal. I am happy and grateful for that. In addition, it gave me the opportunity to participate in two special activities: a Spiritual Life Conference and the Aerospace Career Exploratory Seminar. These, along with five summer encampments, gave me a chance to learn about the Air Force, other people, and myself. In Civil Air Patrol more than anywhere else, I learned more about my attitudes toward achievement, people, and what I wanted to do with my life."

Bob Hickox first became a CAP cadet in April 1967 joining the Albany Composite Squadron. Ultimately, he earned a cadet colonelcy and went on to become a chief warrant officer in the Senior Transition Program and a rated CAP pilot. He holds the Frank Borman Falcon Award, is a member of the Falcon Cadet Squadron and is cadet-in-charge of the Civil Air Patrol Club at the academy. Active as a "disc jockey" with KAFA, the cadet radio station, Hickox also made the Dean's List for Academic Excellence.

Mike Snedeker sums up in a single word what he believes is the "chief advantage of being a CAP cadet before coming to the Air Force Academy."

"Experience," he says, "and the further along a CAP cadet is in the CAP ranks, the better his performance at the academy is likely to be."

"On arrival, Snedeker recalls, "you have to learn basic military skills and Air Force knowledge. Marching, uniform care, customs and courtesies, Air Force aircraft, and many other subjects must be mastered, almost all of which a CAP cadet has already learned.

"Basic summer training demands high physical effort, keeps the new 'doolie' running all the time. The stronger you are as a result of following CAP aerobics, the better off you'll be on the athletic fields.

"During the four academic years, CAP training comes in handy again and again. Whether it is writing a staff study, delivering a briefing, learning navigation, or learning to fly. Prior experience in CAP is what makes the difference between 'average' and 'better than average' performance."

When he became an upperclassman, Snedeker says he fully realized how "invaluable" the experience he got in a position of CAP leadership really was, adding:

"Knowledge of counseling people on performance, as well as how to drill them and teach the basic military skills they need, tends to increase your worth to your squadron and to the Cadet Wing." "Experience", Mike Snedeker concludes, "that's the key, the thing that gives Civil Air Patrol cadets the jump on their contemporaries. I'm grateful to CAP for the experience it gave me prior to becoming an Air Force Academy cadet. Now that I've graduated, I can see where CAP has helped the most."

A former National Capital Wing member, Snedeker joined the CAP program in 1968 with the Col. Virgil I. Grissom Cadet Squadron. He also rose to the grade of cadet lieutenant colonel and command of his squadron before entering the academy where he was active in the formation of the CAP Cadet Club as a "recognized activity" within the Cadet Wing. Like Hickox, he holds the CAP Frank Borman Falcon Award.

The life, experiences and opportunities available to a Civil Air Patrol cadet are not exclusively oriented to the local squadron/community level nor the routine of military drill and aerospace studies. Over the years, a wide variety of "special activities" have been developed. As opposed to attendance at cadet summer encampments held at Air Force bases and other Department of Defense establishments, which are a mandatory part of the cadet program, special activities are voluntary with regard to individual participation. In fact, they are available only to those cadets who meet certain criteria. These criteria vary depending on the exclusivity of the activity, but in all cases (except for the Christian Encounter/Spiritual Life Conferences) the cadet must at least have mastered four achievements in his Phase II training program.

Considered by most cadets the most glamorous of these is the International Air Cadet Exchange (IACE). This also is the oldest of CAP's cadet special activities. It is a month-long adventure in international understanding, goodwill and fellowship. The Civil Air Patrol has exchanged cadets with similar organizations in Canada, Central and South America, Europe and the Middle and Far East. More than 20 friendly foreign nations participate exchanging some 200 young men and women each year. Air transportation to and from the U.S. is provided by the Air Force.

The Cadet Officer School (COS) also is one of CAP's most coveted special cadet activities. Ordinarily held at Maxwell AFB, Ala., home of the Air University, the COS is a two-week course designed to "increase the effectiveness of cadet officers, offering exposure to a curriculum which includes psychology of leadership, problem solving techniques, public speaking, physical fitness and orientation trips. Instruction is divided between lecture and seminar and a field exercise is included. A formal military graduation parade tops off the course.

For those cadets who lean toward the more rugged aspects of Air Force life, there is the Air Force Academy Survival Course (AFASC) held at the academy and in the nearby southern Colorado mountains. This week-long course is planned and conducted by Air Force specialists assigned to the academy and is designed to acquaint the CAP attendees with the art of survival. The course includes water survival, sustenance of life while living off the land and life-sustaining techniques for mountainous country.

Not all special activities necessarily are oriented toward the Air Force and the military. The Federal Aviation Administration offers the FAA Cadet Orientation Program (FAACOP) at its academy, Will Rogers Field, Oklahoma City, Okla. The course provides information on career opportunities in the FAA and with the entrance requirements in addition to acquainting the cadets with the history, organization, function and responsibilities of the FAA and its far-flung elements.

The National Aeronautics and Space Administration also hosts a Space Flight Orientation Course (SFOC) at one of its major development centers, the Marshall Space Flight Center, Huntsville, Ala. This one-week course is intended to further the cadets' aerospace education exposure and to motivate them towards seeking careers in aerospace and allied sciences.

At least three additional Air Force-oriented special activities are offered each summer. They include the Air Training Command Familiarization Course (ATCFC) which provides familiarization training at an ATC Undergraduate Pilot Training base; the Air Force Logistics Command Orientation Program (AFLCOP), a week-long stint at an AFLC base where the cadet begins to understand the breadth and scope of AFLC's global support mission; and the Medical Services Orientation (MSOP), a one-week activity planned and conducted by active USAF medical personnel designed to acquaint cadets not only with the various medical fields operating in the Air Force but also the career opportunities in these fields as they are encountered in civilian life.

Several Christian Encounter/Spiritual Life Conferences (CSC) are held each year, sponsored by USAF chaplains. These are intended to augment the spiritual and moral aspects of the Civil Air Patrol cadet training program and "to stimulate active participation in the church of the cadet's choice."

Finally, there is a National Drill Competition (NDC). Essentially, the NDC is a year-long activity since it begins at the squadron/community level almost immediately after the previous year's NDC winner is selected. Individual cadets compete for places on the unit team. Squadron teams compete for group and/or wing honors, the winner representing that wing in the regional competition. The eight regional teams then meet to determine the national championship drill team.

In all, approximately 900 CAP cadets have opportunities to participate each year in the various special activities while another 6,500 cadets and some 1,000 senior members attend summer encampments at military installations.

Occasionally, other very special opportunities become available to outstanding CAP cadets. One of these appeared in the mid-1950s. Arrangements were made for a Civil Air Patrol cadet to accompany the U.S. mission to the Antarctic. When the huge Air Force Globemaster transports departed Donaldson AFB, S.C., in February 1957 for Operation Deep Freeze, aboard was Cadet Maj. Robert N. Barger of Peoria, Ill. Before Bob Barger's four-month-long adventure ended he became—

. first teenager in the world to fly over the South Pole,

. member of the first U.S. Air Force crew to fly over the South Pole,

. first teenager to celebrate his birthday at the bottom of the world,

. first Catholic youth to serve Mass as an alter boy on the continent of Antarctica,

. one of four volunteers (the others were Air Force and Navy men) to swim in the 29-degree waters of McMurdo Sound, Antarctica, testing new military survival equipment, and

. the only CAP cadet ever to meet in a single day the Secretary of the Air Force; the Air Force Chief of Staff; the vice chief and deputy chief of staff, USAF; the commander of the WAF and the near legendary father of CAP, Gill Robb Wilson.

Today this once-in-a-lifetime adventure is only a memory to Barger. But, when it begins to fade against the challenges of contemporary life in these United States, Bob can pull out a very special letter, one that occupies a special place in his personal treasure chest. The letter, from a co-member of that crew that first put U.S. Air Force wings over the South Pole, reads:

> "Dear Cadet Barger:
>
> "The trip to New Zealand and Antarctica and return, which we have just completed, symbolizes many things to me.

"The fact that our C-124 made this trip without happenstance or incident didn't 'just happen'. It was the result of team work of all crew members and of those who are responsible for its maintenance on the ground. In other words, the results bespeak the ultimate in airmanship and proficiency.

"I want to commend you for your part in this historic flight, which is the first one ever made by an Air Force airplane over the South Pole. The fact that I was able to report that the trip was a routine one further reflects your high degree of proficiency and ability to participate in Eighteenth Air Force operations anywhere in the world.

"My hearty congratulations to you and to the whole team of which I am proud to be a member.

"Sincerely yours,
CHESTER E. McCARTNEY
Major General USAF
Commander, Eighteenth Air Force"

In addition to qualifying for one or more of the special activities during the approximately six years covered by the complete program, the cadet also is eligible for college grants and scholarships and for participation in the CAP "matching funds" program designed to assist him in engaging in actual flight training through civilian facilities.

CAP scholarships are awarded for four-year periods and are renewable annually upon proof of acceptable academic performance. They are not automatically renewable. A cadet, however, may qualify for more than one grant. Special scholarships and grants are made available from time to time either separately or in conjunction with established recurring scholarships and grants and, in addition, other scholarship and grant opportunities are available at the region and wing level in many areas. The majority of the scholarships and grants cover undergraduate, advanced undergraduate and graduate work at accredited colleges and universities, however, one group of grants is made available to cadets interested in furthering their education in special aerospace courses offered by accredited trade, technical and vocational schools.

The fields of study for which Civil Air Patrol scholarships and grants are available must be directly related to aerospace. Recipients must maintain an academic standard of work acceptable to the school concerned and failure to maintain this level of achievement can result in cancellation of the award.

The total number of scholarships available to cadets in any one year is in direct relation to the financial status of the CAP corporation at that time. Since CAP is self-supporting and receives no direct financial aid from any agency of the government, its funds are limited. However, by the mid-1970s the corporation was making more than $40,000 in scholarships and grants to eligible cadets each year. Technical/vocational grants are in the $500 range. Undergraduate scholarship/grants run from $500 to $1,000 while the available graduate grants are for $1,500.

The CAP solo flight scholarship program is administered at the wing level. Each fiscal year, the corporation allocates solo scholarship monies to the wings based on a ratio of wing cadet recruitment to national cadet recruitment. If any wing commander cannot use his funds in a particular year, they are then made available to wings that require additional money.

Cadet flight training may be accomplished either on a continuing basis where the wing contracts with an FAA certified flight instructor or school on the local level for individual cadets or small groups or in centralized encampments at group, wing or regional level. Eligible cadets may work toward their initial FAA power solo license or a glider pilot license. There is a growing trend to accomplish this level of aeronautical proficiency in organized encampments. Glider pilot training is conducted in some wings on an organized continuing basis at the unit level and in other areas on an individual one-time basis.

In all cases, the matching funds program represents a giant step for young men and women who set their sights on a flying career. On the average, to solo at an FAA-approved school costs approximately $360. Under the Civil Air Patrol flight scholarship arrangement, a cadet may solo for one-third of that, since one-third of the cost is defrayed by the corporation and one-third by his wing. Like the scholarship program, however, availability of matching funds depends on the financial health of both the national CAP corporation and the individual wing. In 1973, for instance, $75,000 was made available at the national level. With another $75,000 coming from the individual wings and a like amount contributed by the individual trainees, Civil Air Patrol conducted a cadet flight training effort representing an expenditure of nearly a quarter of a million dollars.

Special activities, college scholarships and flight training, however, are not the life blood of the Civil Air Patrol cadet program. Rather it is the day-to-day exposure to aerospace studies, aviation-related experiences, the character-building aspects of military-style discipline and training, and the intellectually broadening effects of social encounters in nearly 2,000 community-level squadrons and flights across the nation. And it is the untiring work of selfless senior members and inspired cadet leaders that puts it all together.

Each year the Civil Air Patrol selects a single unit as the nation's outstanding squadron. In 1972, the Evanston (Illinois) Cadet Squadron was runner up. In 1973, the Evanston Cadet Squadron got top billing and a $500 cash award for its unit treasury. Its commander, Maj. Sue Sturgeon, a 15-year CAP veteran, was given the F. Ward Reilly Leadership Award.

In the Civil Air Patrol News—the organization's monthly publication distributed to all its more than 60,000 members— Lt. Col. Bill Rechtenwald, Deputy for Cadets, Illinois Wing, explained the "how and why" of the Evanston unit's success. Rechtenwald wrote:

"Evanston, Cadet Squadron of the Year ... no wonder, they have everything going for them, funding, a big senior staff, aircraft and pilots, all of the things we never had.

"Let's set the record straight: Every member pays $1 per month squadron dues; activities are run on a "pay as you go" basis; the squadron has not had an organized fund drive in over five years; the 13 seniors on the roster are thinned out some when you consider two are retired from CAP and six are away from the Evanston area serving with the Air Force; yes, they have some pilots, but no aircraft, in fact, it's a 40-minute, 18-mile drive to the closest airport.

"Well, then they must be doing something right, what is it? Next to the park building where the Evanston Squadron meets there is an elevated railroad that runs to Chicago. One evening last summer the usually docile commuters were startled to see full-uniformed Army Special Forces personnel repelling down the side of the railroad structure as they demonstrated mountain climbing techniques to the cadets of the squadron."

Sue Sturgeon, commander for over two years, explains it this way, "Our whole attitude toward the program has to be a positive one; if we stood around waiting to find a mountain or cliff to use during these classes, we'd still be waiting. We use what we have, but more important, what we have ... we use."

The major, a mother of seven, an avid camper and hoping to start pilot lessons soon, puts a great deal of responsibility on the cadets in her unit.

"For as long as I can remember Evanston has been a cadet squadron," she says, "not just on paper, but in reality. The most important job of a cadet unit is to develop dynamic Americans and aerospace leaders. You don't develop anything by handing cadets everything on a silver platter. I already know how to run a cadet squadron, I don't need practice. What I want to see happen is that the cadet staff, and we have a great one, is able to run this unit. This gives our cadets the leadership experience that no other youth organization has to offer.

"Recruiting is a problem to many units. Evanston might not have this one beat, but it's staying even. A concentrated effort of school presentations, news media coverage, public displays and person-to-person recruiting have been so successful that I recently used the Evanston program as a core for a very successful statewide recruiting program.

"The cadet program works well at Evanston. In that period of time that it became mandatory and all-to-many people stood around talking about how it could never work, the staff at Evanston Cadet Squadron talked about how to make it work better; the squadron has passed more contracts in 1971 and 1972 than any of the other 62 units in Illinois with cadets. A small note on the unit bulletin board reads, 'some men see things the way they are and say why? Some men dream of things the way they should be and say why not?'

It's not all books.

"I remember it was raining on that April afternoon, when I received a call from the cadet commander telling me to wear fatigues to the meeting and bring supplies for three days," reminisces CWO Ed Sackley, Illinois' first Advanced Cadet Transition member and former Evanston cadet commander, "he said there had been a tornado and we were going out to assist in relief."

Within hours 30 fully-equipped Evanston cadets were enroute to the scene of a tornado which hit Belvedere, Ill., killing 25, injuring 500 and inflicting damage of over 25-million dollars.

"That was some of the hardest work I've ever done," Sackley added. "We were loading trucks with debris as we searched for victims. I stopped counting when we got to 50 truck loads. We worked about 18 hours a day for the entire three days we were there. Our CAP training really helped. We were organized. Many of the other volunteers were not. I remember the county sheriff telling us that we did three times as much work as other groups."

The Evanston squadron meets in a suburb just north of Chicago. Because of its proximity to public transportation it has a broad cross section of cadet members which include virtually every socioeconomic, racial and ethnic group. There are former Evanston cadets attending the Air Force Academy, the Naval Academy and the U.S. Military Academy at West Point. In the past, former cadets have also attended the U.S. Merchant Marine Academy. Upwards of 20 cadets have gone on to serve as officers in the armed forces; and numerous persons have joined the enlisted ranks. Thirty percent of the cadets in the squadron have soloed, and this summer six Evanston cadets are taking part in the International Air Cadet Exchange.

"Some people wonder why we always seem to have the lion's share of special activities participants," commented Major Sturgeon, "but what they don't think about is that we have the lion's share of qualified cadets. We'll give our ninetieth Mitchell Award this year. Anyone can do it, but it takes work. As the old saying goes, 'There is no way a person can sit on his tail and skid up hill!'

The Evanston Cadet Squadron is a unit with a good future and a great past. It's not trying to be number one this year, and it didn't try to be number one last year; it just tries to do the best job it can with the people and resources it has. It spends considerably more time pushing itself forward than patting itself on the back. One thing that is most evident in a visit to Evanston is that the people enjoy being there. It's fun, there is a high degree of comradeship and esprit-de-corps.

The Civil Air Patrol cadet program of the 1970s is vastly different from that provided during CAP's formative years and, for the sake of clarification, the term formative years in this context is extended to include most of the 1950s. It was during this period that CAP's formal aerospace education program was established and the study materials developed. Prior to that time, a cadet's training consisted of military drill, discipline and courtesy usually instructed by a veteran of World War II or some of the newly returned Korean War veterans and aviation "ground school" type training provided by the unit's flying personnel or, when the unit was exceptionally fortunate, by a bona fide ground school instructor.

During the 1950s an integrated series of aerospace education texts, each accompanied by a work book and an instructor's guide, were developed by National Headquarters under the supervision of professional educators employed by the Air Force specifically for this purpose. The texts, which regularly were up-dated to keep pace with the advancing aerospace technology, included such titles as Aircraft in Flight, Power for Aircraft, Navigation and the Weather, Airports and Airways, Introduction to Aerospace, Challenge of Aerospace Power and The Dawning Space Age. In late 1974 a new single-volume text entitled Your Aerospace World replaced the seven existing texts used in the Civil Air Patrol cadet program. Texts are made available to the individual cadet at a nominal cost. Available for instructors are complete kits in which the texts, work books and instructor's guides are backed up with audio visual aids including 331/3 rpm recordings, full-color film strips and slide presentations.

On the surface it would seem that with the assistance of the Air Force—in this case the services of full-time professional education specialists and qualified Air Force officers to develop a training program and the materials to support it—that the course of the CAP cadet program would proceed without a ripple. This, however, has not been the case.

There have been growing pains, changes resulting from practical experience in the field, and modifications to meet the changing mood and values of American youth. There have been times when the pendulum has swung too far to one or the other side of center; this with respect to the emphasis placed on the academic portion of the program versus the activity portions and with respect to implementation of the aerospace education portion, specifically with regard to whether it would be more productive as a structured program given as part of the regular unit meeting or whether it should be entirely self-study. In this case, the total responsibility would be placed on the individual cadet to integrate it with his other commitments, establish his own priorities and study at his own pace.

In fact, as the Civil Air Patrol approaches the mid point in its fourth decade the exact nature of the program and the manner of its implementation still is, to some degree, in a state of flux.

The late 1960s and the early 1970s were difficult years for the Civil Air Patrol cadet program. Traditionally one of the strongest segments of CAP's overall effort, the organization encountered significant problems not only with recruiting new cadets, but with retaining those recruited through completion. This was due in a large measure to the general antipathy of the American public toward the military and thus the uniform when articulation against the U.S. involvement in Southeast Asia was at its height. Many also point to changing customs and modes of dress on the part of young men and women—their penchant for the prime symbol of rebellion, long hair—as a factor.

Obviously, CAP's short, military-styled haircut for males and the "off-the-shoulders" style for females is incompatible with the shoulder-length tresses effected by many young men and the waist-length so popular with young women during this period. Also, incompatible was a general tendency toward a new casualness of dress, a tendency that many parents put down as "just plain sloppiness."

In retrospect, however, it appears that antipathy toward the military and resistance to CAP's dress and appearance codes were not as significant as a trend within the Civil Air Patrol toward an imbalance in emphasis between activities and academics. Experienced CAP seniors who have concentrated their major effort on the cadet program over many years explain it this way:

"First we went overboard on the books. Academic achievement in terms of the aerospace education requirements of the program became the single most important thing. This probably wasn't by design but the administration of that part of the program became so complicated and time-consuming the senior members and even the cadet leaders found little or no time to maintain an aggressive program of activities—by activities we mean flight orientation programs, first aid training, participation in civil defense and SAR training, ground rescue, drill teams, field trips to aviation-aerospace facilities. After all, neither the cadet nor the unit got any credit for these things. If you wanted to get credit at all you had to plow through the books and then turn in the paper work."

It wasn't long before CAP's top leadership came to the conclusion that something was awry. Word filtered up the chain of command that activity programs at the unit level were becoming virtually nonexistent. This triggered a major change in implementation philosophy. The aerospace education portion of the program was placed on a self-study basis. New guidance emanated from National Headquarters urging unit commanders to place major emphasis on their activity programs during squadron meetings. It outlined a wide variety of activities, many of which were totally unrelated to the basic thrust of Civil Air Patrol or aerospace. It, in the opinion of many CAP personnel in the field, "gave the impression that they were prohibited from continuing to conduct the academic portion of the program in junction with other facets as part of the weekly meetings!" To others, it appeared to be "activities for the sake of activities" or a "shot gun attempt to find something, anything to keep the cadet coming to the meeting hall."

Again, it didn't take overly long for CAP national leadership and the Air Force professionals at Headquarters, CAP-USAF to realize a disturbing misunderstanding was in the making. New guidance was developed and distributed. It was made clear that the intent was to achieve a workable balance between the academic and activity phases of cadet training and that, in the end analysis, it was up to the individual unit commander and his staff to establish the balance with which their cadets were most comfortable.

In no respect, however, were the requirements to successfully complete the academic phase of the program relaxed. To do so would be inconsistent with the basic objectives established for CAP by its founding fathers and with level of support expended by the Air Force in its role as big brother and counselor.

It is undeniable that the public attitude toward the military during the final months of U.S. involvement in Viet Nam, changing codes of dress and deportment among our teenagers and the internal perturbations in the CAP cadet program between 1968 and 1973 took their toll. As the Civil Air Patrol neared the mid-point in the decade of the '70s, cadet retention continued to be a serious problem. Recruiting, however, improved. Outwardly there were encouraging signs. Young people, by and large, were turning back toward traditional American values and principles. Except for the dyed-in-the-wool nonconformists, young men and women also began to effect more conservative modes of dress. The differences between contemporary dress standards and those required of a military-oriented organization continued to narrow. You might say that more and more young Americans were taking up Cliff Cushman's challenge.

Simultaneously, Civil Air Patrol leadership began taking steps to resolve the problem of program imbalance. Under the leadership of General Westberg and General Patterson, renewed attention has been directed toward insuring that cadets are afforded greater opportunity for participation in the traditional operational missions of the Civil Air Patrol. Establishment of composite squadrons—those having both active cadet and senior elements—was urged. In such units there are a sufficient number of senior members to more effectively counsel cadets and provide the opportunity for participation in action oriented activities relevant to CAP's basic objectives. In the national evaluation criteria, new weight was placed on providing regular orientation flights for cadets and more importance was placed on cadets completing the orientation full program and achieving the Spaatz Award as opposed to awarding maximum recognition on the basis of in-step accomplishment of the 15 achievements that make up the cadet program's four phases—the beginning phase, the learning phase, the leadership phase and the executive phase.

Finally, the underlying weakness in the program was pinpointed—the quality of leadership at the unit level. An allout effort was initiated to bring into the Civil Air Patrol a sufficient number of men and women who at one and the same time represented the degree of motivation necessary as well as those special attributes which make a leader.

At last report that effort already had begun paying dividends.

Military drill, discipline and courtesy are combined with moral leadership, physical fitness and aerospace education in the Civil Air Patrol cadet program.

"Dress right, dress!" is the command. The voice may be distinctly feminine but nonetheless authoritative in today's Civil Air Patrol. This female cadet officer is one of the top-ranking cadets of her wing and can bark out the commands with the best of the boys.

How do you conduct mountain rescue training without going to the mountains? These CAP cadets in a western state solved the problem with a rig of their own design. A local television station found the rig and the training interesting enough for a special feature.

The graduating class from a Civil Air Patrol Flying Encampment poses for its portrait while four of its instructors make a low-level formation pass down the runway. CAP usually contracts with local fixed base operators for both aircraft and Certified Flight Instructors for such encampments.

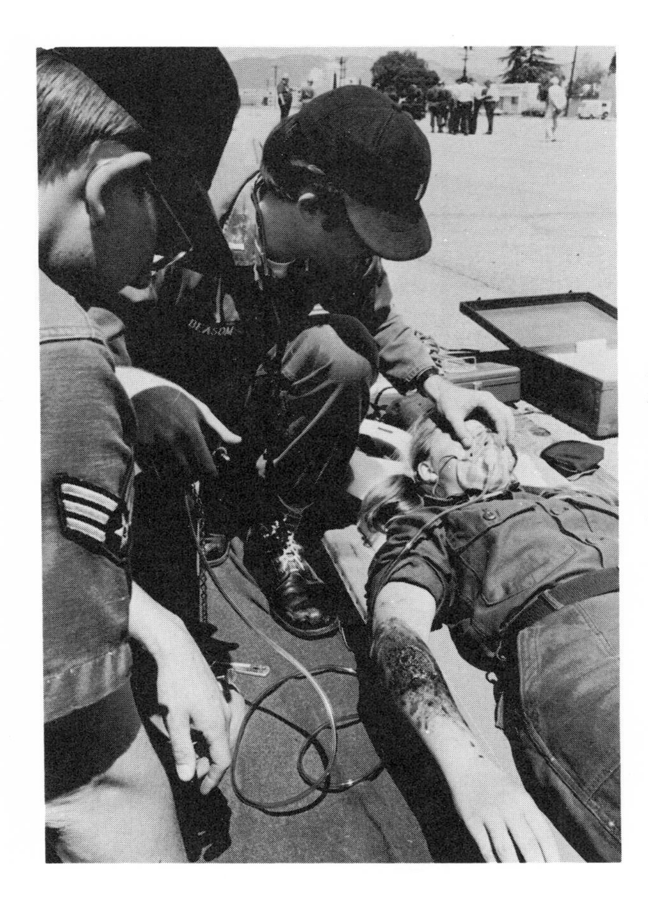

Disaster relief has become one of the Civil Air Patrol's more important operational missions. CAP is attracting many trained medical technicians or paramedics as they are called and many units have aggressive programs to upgrade their first aid and medical assistance capability.

Senior commander of a cadet unit checks out his "men" prior to embarking on an overnight, mountain bivouac. Wilderness survival is included among the many special types of training available to Civil Air Patrol cadets.

Under the watchful eye of an Air Force instructor, these CAP cadets attending the Communications Electronics Course at Keesler AFB., Miss., practice receiving and sending messages.

Circa 1970 CAP cadets get a look at an old-timer (the plane, not the pilot). This Fairchild 24 once flew on anti-submarine patrol with CAP during World War II. Today it is a cherished "classic" aircraft but its owner, a CAP senior member, still flies it on operational search and rescue missions.

"Volunteer" citizen plays victim for CAP ground rescue team during public display of Civil Air Patrol equipment and resources. Victim is securely fastened in "stokes" litter used for removing badly injured persons from difficult locations.

Although CAP cadets do not participate in "high risk" operational missions, they do get a taste of the real thing through regular training exercises. These cadets assist a rescue technician of the Air Force Rescue and Recovery Service, loading a "victim" into an ARRS chopper for evacuation.

The brass listens! Brig. Gen. Leslie J. Westberg, the Air Force general appointed by the Secretary of the Air Force to oversee its civilian auxiliary, meets with a group of the leading Civil Air Patrol cadets to get their perspective on a new cadet program under consideration.

A hard day of training over, these CAP cadets "chow down" in the dining hall at an Air Force base. Cadets learn about the life of regular airmen and officers during their annual CAP summer encampments at USAF installations.

Equal opportunity—equal responsibility has been the hallmark of the Civil Air Patrol since it was formed just prior to World War II. Today female senior members and cadets are well represented in the ranks of CAP's outstanding units. This group commander and her Air Force Reserve Assistance Officer conduct an inspection of her command.

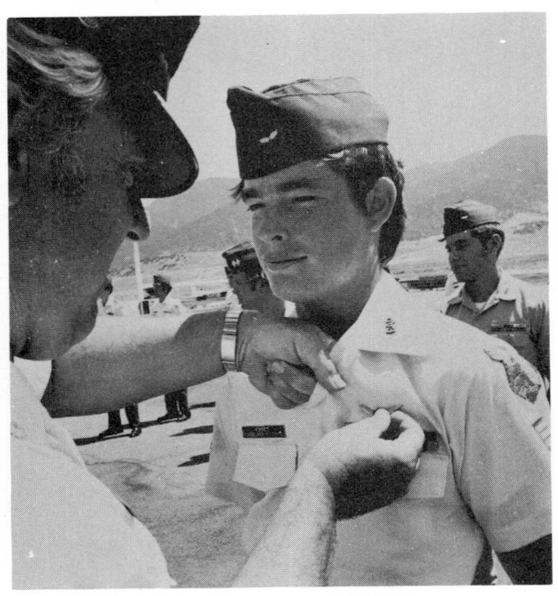

It's a proud day for this Civil Air Patrol cadet as he receives his CAP Solo Pilot wings from the commandant of a CAP Flying Encampment. Cadets may earn their wings at one third the going rate through the CAP Flight Scholarship Program.

Two staff officers watch while a CAP cadet officer puts a new recruit through the paces. Under the watchful supervision of senior members, training of new cadets is largely carried on by older cadets who have achieved noncommissioned officer or officer status.

One of CAP's "most wanted" special summer activities is the Space Orientation Course sponsored by the National Aeronautics and Space Administration. These cadets try out the engineering model of the lunar rover vehicle.

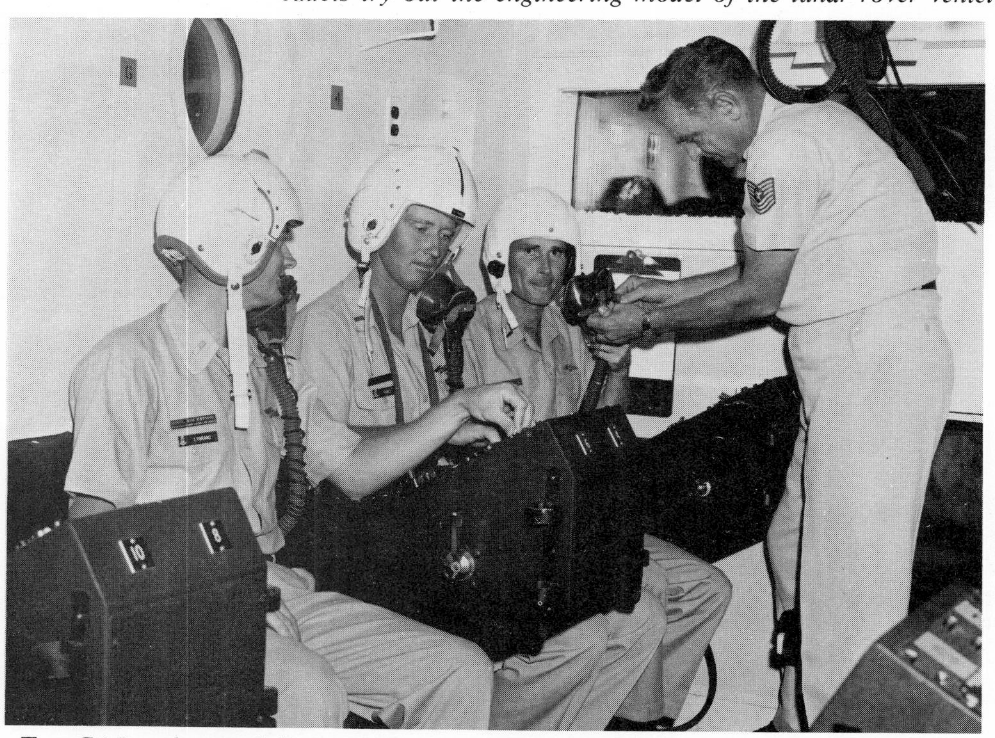

Two CAP cadets and their Air Force advisor prepare for a "flight" in the altitude chamber during a visit to a nearby Air Force Base.

On into the jet age, CAP cadets visiting an Air Force base learn about the Lockheed T-33 jet trainer. While other teenage aviation buffs look longingly from afar at new Air Force, Navy, Army and Marine Corps aircraft, CAP cadets usually get an early and first hand introduction as new equipment comes into the military inventory.

While much of the active Air Force still was waiting to get its look at its new McDonnell-Douglas F-15 Eagle, these CAP cadets were getting a personal introduction to the new air superiority fighter at Edwards AFB, Calif. Visits to USAF installations are a regular part of the cadet program.

During a visit to the local airport these CAP cadets get a close-up of one of the latest civil aircraft to come off the drawing boards, North American's single-engine Aero Commander.

CHAPTER 13

Getting Out The Word

One of the three major objectives of the Civil Air Patrol as set forth in the Public Laws which gave it legal status and also made it an auxiliary of the Air Force is to "provide aviation education and training especially to its senior and cadet members" and to "encourage and foster civil aviation in local communities."

With the advent of man's penetration of space, the term has been amended to "aerospace education"—aerospace meaning a combination of aeronautics (aviation) and astronautics (space). CAP's intensive efforts in this field, begun in the 1950s and, at that time, exclusively embracing aviation as it was understood in the pre-space era, continually are brought up to date reflecting our advancing space technology. It is important to note, however, that the major portion of its aerospace education program still is devoted to the "down-to-earth" aspects of aerospace—general, commercial and military aviation and their supporting technologies.

It is also important to note that aerospace education in the context of CAP's program is designed not to train pilots and technicians but to impart general understanding of aerospace with emphasis on the impact aviation/space and the supporting sciences have on our contemporary way of life; their impact on domestic as well as international politics, the economy, business and industry, leisure and social standards.

Aerospace—man's involvement in utilizing the atmosphere and space in which to travel and perform useful tasks—has been with us only a little more than 70 years. During its first 50 years most of the men and women who became involved in aviation talked primarily to one another. The glamour, excitement, adventure represented by the first 50 years of flight made such an impact on them they could not comprehend that the rest of the world was largely standing by and watching them with an amazing lack of understanding. They knew what aviation would ultimately mean to the nation and the world and assumed everyone else did.

When the United States was "caught with its flying britches down" in the period when it became apparent a second world war was in the making, many aviation leaders made a solemn promise to themselves that "if and when we got out of this one" they would not let it happen again.

Starting from behind Germany, Japan, Great Britain and even France, America's industrial might and technological precocity moved it into a position of world dominance in aviation not only in terms of military air supremacy, but also in commercial application of the airplane by the time the Axis bowed in defeat. Aviation-wise, the United States came out of World War II "smelling like a rose" and a large number of its leaders, among them many who in the dreadful days leading up to Pearl Harbor abruptly recognized the need to spread the aviation word among the public, now relaxed letting the euphoria of success lull them again into inaction.

"After World War II, we thought we would see a great acceptance of aviation not only with respect to the general public, but more important, as a means of motivating young people," Gene Kropf, one of the nation's foremost disciples of aviation, recalls.

Kropf, who became the first director of the Aeronautical Administration Department at Parks Air College, St. Louis University, in 1945, now is director of Public Affairs for the Federal Aviation Administration's Western Region. He is a member of CAP's National Aerospace Educational Advisory Committee and president of the International Aviation Fraternity, Alpha Eta Rho.

"We also thought," Kropf continues, "that a lot of educators, teachers, members of school boards who had been intimately exposed to aviation in the military would come home all fired up to get aviation off-the-ground in the educational system. Nothing could have been further from the fact.

Most of them came back with a belly full of war and war-related experiences. Since the airplane was so closely associated with that particular war it was banished from their minds."

Veteran aviation instructors like Kropf who had looked for sympathetic support from the academic community in introduction of aviation-oriented subjects at the elementary and high school level found not only a lack of support, but in many instances stubborn resistance.

"We thought we would be dealing with a new breed of cat," Kropf declares, "but we faced the same stodgy types we encountered before the war. Even when we found an inspired teacher, there always was the obstacle of the principal, superintendent or school board member who took the position that 'aviation subject matter can not be included in the curriculum without adding similar material on railroads, boats, automobiles and the like'. Academicians and administrators all along the line dug in their heels and motivated teachers found themselves stymied."

The Civil Air Patrol's aerospace education program consisting of professionally-written texts, instructor's guides, student workbooks, long-play records, film strips and 35 millimeter slide presentations all carefully integrated so as to provide a well-balanced course, was first developed in the mid-1950s. Initially intended for the CAP cadet program, it became apparent that, with little augmentation, it could well provide what then was almost entirely lacking—a general course of instruction in aviation-oriented subject matter written by educators for educators. One of the main obstacles to achieving any real progress in general acceptance of aerospace education by the academic world was a dearth of academically-sound, inexpensive and easy-to-come-by materials. The CAP course appeared to fit the bill and Civil Air

Patrol with full Air Force backing launched a major effort aimed at getting aerospace education—through its course—into the nation's schools. Complementing the educational resource materials it was making available at nominal cost, CAP also launched a massive effort to develop (and motivate other organizations to develop) aerospace education workshops for teachers in cooperation with as many colleges and universities as possible across the country.

CAP's circa-1970 aerospace education program is given the same priority as its other two missions—emergency services and cadet program—and, in one respect, is an integral part of the cadet program. It is structured to meet a two-fold requirement: (1) provide a comprehensive self-study course for the CAP cadet—it is one of four curricular factors; and (2) make available to the general citizenry an aerospace education program by which the public can gain sufficient knowledge about aerospace to make "informed decisions" on aerospace issues. Thus, it consists of an internal and an external program.

The external program is defined by CAP as "that portion of the overall aerospace education mission which is concerned exclusively with contacting the general public relative to aerospace matters." It has three component parts: CAP aerospace education school programs; CAP aerospace education workshops; and liaison with the educational community.

The CAP school program makes available to interested junior and senior high schools a one-year, general education course in aerospace education. It is designed to develop in the student appropriate knowledge, skills and attitudes about aerospace activities and the total impact of air and space vehicles upon society.

There are three types of school programs available. All three are suitable for either junior high or senior high school level. They include the CAP aerospace education elective course, the squadron associated program, and the CAP high school squadron.

The one year aerospace education elective course is offered by high schools which use CAP materials and follow the CAP syllabus. This type of class is subject to the requirements of the school but may be used for CAP credit under certain conditions. Members of the class need not be CAP cadets. The squadron associated program is similar to the aerospace elective course except that some of the members of the class are also members of the local CAP squadron. The completion of the high school elective will satisfy the aerospace education requirements for Phases I and II. Cadets must complete their other phase requirements through the local squadron. In CAP high school squadrons all aspects of the cadet program are held or sponsored by the school. Normally, the aerospace education requirement is satisfied through classes formally held at the school whereas other curricular requirements are accomplished in extra class time approved by the school officials. In all cases the program is under the direct control and supervision of the school authorities.

In the early days, the same men who were instrumental in the development of CAP's aerospace education program became active in the National Aerospace Education Council (NAEC) and later the National Aviation Education Association (NAEA). In these organizations they worked as long and diligently in the interests of aerospace education as they did in CAP. In fact, before long, they assumed a dominant role which continued until early in the 1970s. Relatively speaking, their progress was enormous. In terms of what needed to be done that progress barely scratched the surface. During this early period the major general aviation aircraft builders and the big aerospace industries began to generate a quantity of educational material and made it available to both schools and the general public. These materials, however, initially were keyed to sales and by and large were unacceptable to educators. It was not until the 1960s that industry began aerospace education in earnest.

Large commercial and general aviation aircraft manufacturers began employing educators both to develop materials and also to provide an interface with the academic world. Instead of "selling airplanes," their materials began treating the inter-relationships between aviation and the everyday problems of making a living and raising a family. The airlines also got on the band wagon. American, United and Trans World put out the first airline-developed aerospace educational material aimed directly at the schools. Douglas, North American, Boeing led for the big military and commercial airplane manufacturers while Cessna, Beech and Piper represented the general aviation aircraft interests.

With the advent of man-in-space—the National Aeronautics and Space Administration's Mercury, Gemini and Apollo programs—there was nearly a hundred fold increase in the amount of professionally-prepared educational material available to the schools and the public. NASA mounted a 10-year, multi-million-dollar effort to promote public support for the nation's space efforts. This effort, while it achieved its end objective—that of mustering wide public support for the manned space program (thus insuring the necessary funds from Congress)—did little to help the overall advancement of aerospace education, especially that part of it that dealt with machines that operate within the atmosphere.

Over a period of nearly 20 years, CAP's efforts in aerospace education have met with considerable success. An estimated 15,000 teachers a year—for a grand total of approximately 300,000 through the mid-1970s—have been exposed to "curriculum enrichment in aviation/space related subject matter" through the medium of these workshops alone. Hundreds of additional teachers have been afforded similar opportunities during workshops sponsored by other organizations as well as those initiated individually by colleges and universities. This is not to give the impression that nearly a half million teachers immediately returned to their classrooms and began conducting courses in aviation/space subjects. At best most of them were barely able to begin relating aviation/space developments to whatever subjects they were teaching. Many left their workshop inspired and motivated only to come up against a brick wall when they got home to their individual school system. A few found their superiors to be enlightened with respect to the need to inform their students about the developments in aviation and space. These met with resounding success. They were a small minority and even 20 years later, CAP can single out only some 1,600 schools in the United States where its courses—or even portions of its material—are being used.

One relatively recent development is providing some encouragement. Finally, after many years of steady pressure, colleges and universities are beginning to give graduate credit to teachers for their summer study/workshop participation in the aerospace education. It is a way of life for the teacher who wants to advance that he devote a significant portion of his annual vacation period to study and travel for which he can obtain credit toward an advanced degree. It is only in recent years that such credit generally has been extended for aerospace education activities. This is resulting in a significant increase in teacher interest and participation in aerospace education summer workshops. This also is an indication that aerospace education is beginning to win some degree of acceptance at the upper levels of the academic hierarchy.

Gene Kropf observes: "Maybe we are finally on our way!"

One advantage the aerospace education office at CAP National Headquarters has over most of the other staff offices is the fact that it has enjoyed greater continuity of direction over the years. While active duty Air Force officers and airmen assigned to Hq., CAP-USAF normally have only a three-year tour of duty before being transferred to another USAF activity, the aerospace education office has been primarily staffed with Department of the Air Force civilians since its inception. Its original head was Dr. Mervin K. Strickler, who now is special assistant for Aviation Education to the FAA administrator. Strickler and CAP's current deputy chief of staff for Aerospace Education/Cadet Program, John V. Sorenson, keep in close touch. Sorenson, in fact, was a Strickler protege nearly 20 years ago and when he was one of CAP's eight regional directors of Aerospace Education.

Another of the eight regional directors is L. D. "Pat" Cody who, for more than a decade, has held the responsibility for CAP's Pacific Region. This region, because of the logistics involved— it extends from Nevada to Hawaii and from California to Alaska—has always been one of the most difficult in which to effectively implement the program. According to Cody, CAP's aerospace education program in his region is "on the move."

"For the first time," he says, "colleges and universities in the western states are getting so many applications to attend their aerospace education workshops, seminars and teacher orientation courses, some people are being turned away. In 1973, the number of junior and senior high schools providing aerospace education courses doubled. It is now more than 300. I see the number doubling again in 1974 and continuing to increase thereafter."

"At least 80 percent of these schools," Cody adds, "are using all or part of the aerospace education materials made available by the Civil Air Patrol."

A significant problem Cody sees—and he feels it is common to all CAP's eight regions—is that there are not enough people available in the Civil Air Patrol program to get the job done. The Air Force is doing its share providing the national headquarters staff together with the eight regional directors. It is on the CAP corporation side of the fence that the effort "is falling down."

"Frankly," Cody observes, "many wing commanders are not nearly forceful enough in their endorsement of the program. They do not insure that sufficient emphasis is placed on the program at the group and squadron level. They make little or no effort to recruit interested qualified educators on the wing staff to provide the guidance and leadership needed at the community level."

Cody admits that a few wing commanders have, over the years, carried on outstanding aerospace education programs citing Puerto Rico's Col. Clara Livingston as one, but in the aggregate he feels that most CAP wing commanders have been "luke warm to negative" about the program singling out his own region as an example.

"In the 12 years I have had the aerospace education responsibility in this region," he declares, "I have never seen a wing commander—and there have been quite a number in our six states—participate, except in a token manner, in a workshop."

Most aviation authorities today rank the Civil Air Patrol aerospace education program as "excellent." Gene Kropf, from his vantage point with the FAA, points our that it "is one of the few that can show a history of continuity growth and improvement over the years."

"One of its advantages," Kropf says, "is the fact that it has been structured by experts drawn from the academic world and then cycled through the aerospace world. It makes a lot more sense to make aerospace experts out of educators than to try to make educators out of pilots, technicians and engineers.

The CAP program continues to make significant progress. Admittedly, that progress is slow, but it has been no slower than the aerospace education efforts of other organizations and aviation entities. And, as Gene Kropf puts it, "it is still growing." That growth is particularly evident in its increased depth and content.

In recent years the basic texts, instructor's guides and student workbooks have been augmented with complementary material. Programmed learning exercises in subject areas like "How to Study." "Aerospace Education Defined" and "Job Analysis Training" have been developed. A curriculum outline for a college-level aerospace education course has been completed and made available to interested institutions of higher learning. A significant addition to the CAP aerospace education library is a comprehensive Aerospace Education Course Syllabus—a guide for an elective, two-semester (one-year) high school course designed specifically to get the high school teacher over the hurdles of introducing aerospace subjects for the first time. It negates one fairly common criticism of the CAP course in the past, namely that the teacher who is a neophyte to aerospace oftentimes is overwhelmed by the subject matter even though, in the CAP course, it has been significantly simplified and wherever possible made non-technical.

The syllabus takes aviation/space from the "legendary contributions to flight" of the Emperor Shun and King Bladud to the real-life, real-time exploits of U. S. and Soviet astronauts in space. Finally, the professional aerospace educators employed by the Air Force to staff the appropriate activities at CAP National Headquarters are geared to provide a wide variety of specialized services in the interests of advancing aerospace education. These include curriculum consultation, administrative guidance, resource identification, guest lecturers, professional consultants, field trips, orientation flights for teachers and briefing teams. Of course, these specialized services are available only to the extent that funding, personnel and airlift are available. Ultimately, that question revolves about the question of national and Air Force priorities.

In the final analysis, it can be said that the Civil Air Patrol has made a substantial contribution in the past, "getting out the word" through aerospace education (it was one of the pioneers in this field); it continues to make a significant contribution today; and it stands a good chance of making a similar contribution in the future.

Of one fact, there can be no question. The need for aerospace education is as critical today as it was in the days of Billy Mitchell. From a purely parochial point of view the resounding defeat of the U.S. supersonic transport development program at the hands of the Congress left no doubt of that.

A veteran educator, Dr. L. Rogers Liddle, associate dean of Education at California State University at Los Angeles, who is a newcomer to both aerospace education and the Civil Air Patrol, puts it in larger perspective this way:

"Aerospace education can well be a question of national survival."

Liddle, who in three short years admits he has become an "aerospace education activist," sees space and its conquest becoming the "common antagonist for the forces of the United States, the Soviet Union and the People's Republic of China." Therein may lie the future salvation of mankind.

"In the normal progression of events all through history," Liddle observes, "it has been the nature of man to fight. It is a continual process of life. In prehistoric times, they fought one another. In Biblical times, banded into tribes, they fought tribe-against-tribe. During the feudal period, tribes banded together and fought one another in larger groups. These groups developed into nations. The fight—nation against nation—went on. Today the countries of the world are banded together into groups of nations. It follows, therefore, that these groups of nations must, sooner or later, battle it out in that final war that will end civilization as we know it.

"Unless—unless they, faced with a common antagonist, can unify seeking a common goal. The conquest of space can be that goal; the dangers of space the catalystic antagonist; peace among men the ultimate achievement—that achievement reached not by the direct approach, the end of fighting among men because that is not the nature of man, but by indirection, the assault on common antagonist.

"Perhaps, that is what we hope to achieve through aerospace education."

CHAPTER 14

CAP Is A Big Business

How well does a far-flung nationwide organization staffed entirely by volunteers on a part time basis manage a multi-million dollar operation?

In the opinion of the Air Force professionals assigned to look over CAP's shoulder, "pretty darned well!"

"We are amazed," they admit candidly, "at just how well it works."

While these kind words reflect considerable credit on the rank and file as well as on the command structure of the Civil Air Patrol, they are particularly welcome to a handful of members—those members who take on and perform the wide range of "dirty, thankless, usually frustrating, behind-the-scenes" jobs necessary to keep CAP financially solvent and supplied with everything from airplanes to paper clips vital to accomplishment of the mission.

"Aircrews, ground teams and communicators live the 'glamorous life' and get the credit for saving lives. Cadets get the applause as they march, heads held high, in the local parade. And the aerospace educators enjoy the 'mutual admiration society' of the academic community."

At least that's the way most of the "pencil pushers" of Finance and the "dock wallopers" of Logistics see it.

There must, however, be some hidden, secret reward for performing these inglorious and tedious jobs in CAP, because the clerks, accountants, CPA's, mechanics, warehousemen, machinists, welders, technicians and just plain "go-fers" who keep CAP on the road and in the air do one hell of a job and emerge as perhaps the most highly motivated of all the men and women wearing the uniform of the USAF auxiliary.

Perhaps one reason for this is that it appears by-and-large these men and women are doing what they know best and in more instances than not their CAP assignment is but an extension of the trade or profession by which they earn a living. Their satisfaction comes from a balanced set of books, a good unit bank account, a smoothly operating airplane or a sharp, Air Force-blue vehicle bearing the CAP seal.

This is not to say that the Civil Air Patrol has enough of these people. It hasn't, not by a long shot. Many units barely keep their heads above water with respect to properly managing their resources. Recruiting more financial and materiel managers, particularly at the squadron and group level, has a high priority because, although by law the Civil Air Patrol is a non-profit corporation, it is big business, a business that must effectively manage annual expenditures of nearly $1,000,000; manage and properly account for nearly $900,000 in state appropriations made to 31 of CAP's 52 wings; and manage, maintain and operate an aircraft fleet with an estimated value of $8 million and a vehicle fleet having a replacement value several times that figure.

This also is not to give the impression that CAP is fat. It is not. It's multi-million dollar inventory of aircraft, vehicles and communications equipment largely consisting of material "excess to the needs of the Department of Defense" given to Civil Air Patrol through the Air Force consistent with the provisions of Public Law 557.

In the cold light of the real world, those words "excess to the needs" are misleading. Almost without exception, by the time these resources are declared excess, the service responsible is glad to move them out. The property disposition system within DOD goes something like this. When vehicles or equipment are found to be surplus to the needs of a military unit, all other units within that service get first opportunity to acquire it. If no takers are found in that service, the items then are made available to the other services. If no further use is indicated by any of the services or independent DOD agencies, the CAP and organizations like the Army, Navy and Air Force Military Affiliate Radio System (MARS), a volunteer organization of licensed radio amateurs who provide back-up communications, get what is left.

It is true that sometimes the equipment, particularly electronics test equipment and similar devices, are relatively new. But, in these cases the equipment either is technologically obsolete or is too highly specialized. In the final analysis, when vehicles, aircraft and other material become available to CAP, these gifts carry with them an automatic obligation to spend large amounts of money either to make them serviceable or, as in the case of DOD aircraft, make them meet current Federal Aviation Administration regulations so that they legally can be operated in performance of the CAP mission.

A case in point—the 38 Cessna T-41 aircraft made available to CAP in 1972. To convert these aircraft for CAP use (and to overhaul them so that the recipient unit would be receiving a new resource not a white elephant) the corporation invested $147,835 or an average of $3,890 per airplane. Ultimately, the individual wings which received the T-41s had to reimburse the national treasury. Some wings received aircraft with "zero time" engines. Where the engine time was low enough complete engine overhaul was not required, the wing's level of reimbursement was altered downward accordingly. In other words, an "engine credit" was given reducing the wing's cash outlay.

As one CAP wing commander is wont to say: "Anyone who thinks CAP gets anything for nothing has rocks in his head."

In the early 1950s, the Civil Air Patrol enjoyed a relatively "fat" period with regard to aircraft as well as to certain vehicles and equipment made available by the Air Force. Such items were placed "on loan" to CAP. This meant the Air Force retained accountability for them, and within its capability to do so, provided maintenance. While this system had distinct advantages for CAP, it was not advantageous to the Air Force, particularly in terms of limiting direct USAF outlay in support of CAP. It was determined that such items, in the future, would be given outright to CAP with "certain strings" attached. First, none of these items may be disposed of by the Civil Air Patrol without a written release from Civil Air Patrol-USAF, the Air Force side of the CAP national headquarters house (granting such release is strictly governed by both USAF and CAP regulations) and, second, any proceeds from such authorized disposal accrue to the Civil Air Patrol corporation (National Headquarters) and are disbursed back to the wings in accordance with established procedures.

With elimination of the "on-loan" program and the advent of acquisition of this property by the CAP as "gifts", the CAP membership collectively and individually assumed the total financial responsibility for making them usable and for keeping them in the organization during the mid-1950s. Suddenly, almost overnight, it seemed to most members the "on-loan" aircraft—several hundred, L-4s, L-16s, L-5s—with which they had been doing a magnificent job and representing their pride and joy, became as millstones around their necks. In cases it was weeks or months before these aircraft had valid FAA airworthiness certificates and legally could be put back into the air. Some were grounded for a year or more. A few never got off the ground again.

The situation that resulted did, however, blast CAP out of a rut of dependency into which it had fallen. From the bottom to the top it became evident to CAP leadership that it not only had an operational mission to perform, now it had a real brute of a management task ahead. It was during this tumultuous period that CAP met this new challenge and took another giant step toward corporate maturity.

Today, comprehensive procedures outlined in voluminous manuals and regulations provide specific guidance to all echelons of command for management of both financial and real property resources. CAP's financial and materiel management systems closely follow those used by the Air Force with departures only where compliance with civil statutes and contemporary business practices so dictate.

In the aggregate, CAP does well in the management of "high value" items, but in the opinion of not only the Air Force but also veteran CAP Logistic officers, "management of low value items doesn't go so well."

One of those CAP Logistic officers is LTC Robert N. Walker who in civilian life has some 20 years experience in managing millions of dollars worth of government property being used by major contractors in the aerospace industry. Walker, for instance, wrote the property management manual for the Northrop Corporation, one of the nation's largest aircraft manufacturers, and currently is a contract administrator for Rockwell International's Space Division. In other words, he is a pro. You might also say Walker is a professional CAP officer since he has been a member for some 22 years; has been in Logistics for seven; and, through involvement at all levels, is intimately familiar not only with the requirements of CAP units at all levels but also with the problems associated with filling those requirements.

Walker has some very specific thoughts on the management of CAP resources. One of these is the fact that "CAP's financial and materiel resources are inexorably entwined" and that a closer rapport must exist between Finance and Logistic officers at all echelons. He also sees the need for unit commanders to be "more informed in terms of resource management and become more directly involved in seeing that they have personnel assigned who are willing and have the capability to perform these critical functions."

He points out with some emphasis that lack of understanding or support of these functions isn't limited to just commanders at lower echelons.

"Some wing commanders," he says, "don't seem to fully comprehend the fact that as officers of the corporation they are held legally responsible for the proper management of CAP resources. I've served several wing commanders during my tenure with CAP and, frankly, some of them didn't realize just how legally vulnerable they were. I have, at times, been criticized for being too "hard-assed" in the manner in which I performed my job in Logistics. That's when I pointed out to them that I had to be hard-assed to save theirs."

"The blue suit Air Force," Walker declares, "is a big business today and it is run like a big business. CAP is a big business and must be run the same way."

On the subject of Finance and Logistic officers and their role in CAP, Walker uses the analogy of the weekly poker player.

"If he doesn't show," he says, "it just leaves a hole at the table. The game goes on. If the Finance or Logistic officer doesn't show up for CAP meetings or work sessions, there is a hole in the whole unit's performance."

Perhaps the biggest problem in the area of Logistics—it apparently isn't quite as serious in Finance—presents itself with change of command and in an organization of volunteers, change of command can be almost a weekly occurrence until you find just the right man or woman.

"When a new commander takes over a squadron," Walker emphasizes, "it is imperative that he inventories the accountable materiel for which he is assuming responsibility. For his inventory to be meaningful, it also is necessary for the departing commander to have taken a final inventory. Only by comparing the two inventories can the new commander be assured that, in fact, the items charged out to that unit on group/wing records are on hand. The new commander must bear in mind he is responsible whether or not the previous commander maintained adequate control and accountability.

"Sooner or later, higher headquarters is going to audit the unit's materiel accounts. This becomes one devil of a job after you churn the pot through several changes of command. Sometimes it becomes impossible to trace certain items of equipment with the result that individuals involved find their personal integrity besmirched both within and outside the CAP. All this can be avoided if commanders just take time to maintain a property inventory that is accurate at all times.

And with the few items of accountable property usually in the possession of a squadron this is no major task."

Obviously, Walker's views on the subject are widely concurred in at the national management levels.

One of the major problems of CAP—and it is a problem both to finance as well as to supply people—is the cost involved in storing, handling preparing for distribution and distributing to units materiel coming from DOD channels. Once Civil Air Patrol has taken possession of the items, it becomes CAP's responsibility to accomplish further distribution. This is expensive.

In most areas, a system has been established whereby the gaining unit is assessed a nominal amount to defray out-of-pocket costs incurred in completing both the paperwork and the physical effort involved. This can amount to a few cents on expendable materiel or a few dollars on small items of equipment, but, can run into the $50-to-$100-range when vehicles, for instance, must be accepted by CAP at the DOD storage point and transported several hundred miles to the area where they are to be used. In the final analysis, these monies come, more often than not, from the pockets of individual CAP members. It is in this area that CAP faces one of its major difficulties.

Take, for instance, a half-ton truck. The vehicle, along with a number of others, becomes available at a DOD property disposal point in the mid-west. There are sufficient vehicles to meet the requirements of several wings. For units of the wing in which the disposal site is situated, the cost of moving its vehicles is relatively minor. For nearby wings the cost increases with the distance to be traveled. If the vehicles are in average shape—considered economically repairable in military terminology—many of them still will require some maintenance before being moved. Additional costs now have been incurred for which the wing or other CAP entity assuming responsibility for the distribution must be reimbursed.

Once the vehicles have arrived at the respective wing headquarters or supply depots they can be issued to units having a requirement. Procedures vary with the wing. In some areas CAP through its own capabilities, refurbishes as necessary, paints and decals the vehicle, then charges the gaining unit for the service. In others, the bare-bones vehicle is turned directly over to the unit or wing staff agency having a requirement with the understanding that the responsible unit or individual "immediately" perform such work as is necessary to bring the vehicle up to snuff. In either case, a substantial amount of individual member funds is involved. Sometimes a single member bears the cost, other times small groups of members share them.

This is where the rub comes in. This member investment is time and money. The member/members think they have a vested interest in the vehicle (that interest becomes substantially larger in the case of an aircraft). Notwithstanding, the legality of the situation, those members then feel they "own" the property. To them it becomes their vehicle or airplane. This proprietary interest becomes more acute on the part of the individual if his contribution was not made in the proper spirit.

Where the individual paid out the cost taking it as a legitimate donation to CAP on his Federal income tax and where his donation was made in the spirit of helping CAP perform the mission, there is no problem. But, where the individual funded the project to enhance his own use of the equipment or for the purpose of personal prestige, he usually takes the position that it is "his vehicle" or "his unit's vehicle."

Immediately the "clique syndrome" manifests itself. The aircraft, vehicle or other equipment is not made available for use by "outside" CAP members even though they are properly qualified to operate it. In the case of airplanes— both those obtained from DOD and those purchased by CAP units—the mode of operation becomes one of the "flying club." In other words, if an individual hasn't contributed to the purchase or the overhaul of the aircraft, he is actively discouraged, if not "shut out", from flying it. This not only is in direct violation of both the spirit of CAP and of current regulations, but it also has a severely detrimental effect on aircraft utilization and further tends to seriously limit the cadet flight orientation effort.

In 1973, the CAP National Board, in cooperation with CAP-USAF, took action to curb this tendency. The commander's evaluation—a National Headquarters management tool—was amended to quadruple the points a wing can be awarded for aircraft management. This placed wing commanders under more pressure from regional and national levels to effectively utilize corporate aircraft. It made it necessary—for the commander who wanted his wing to improve its rating in competition with the other wings—to reassign or even dispose of aircraft which were not utilized effectively. Threat of reassignment already is resulting in an improvement in the aircraft utilization rate. In a companion action, the evaluation criteria were again changed in 1974 to increase the number of points awarded for cadet orientation flights. CAP's national leaders also have made a point of emphasizing to wings, groups, squadrons and individuals that "all aircraft, vehicles and equipment acquired by CAP by donation, gift or purchase become the property of the corporation and not the property of any individual member or unit" and as such are subject to be moved or reassigned as necessary in the effective performance of the CAP mission.

Air Force and CAP corporate leadership also point out that "any individual who has CAP assets is responsible for the proper utilization, control and protection of these assets." They are to be used exclusively in the furtherance of CAP missions and "may not be converted to personal use." Such conversion, of course, is a violation of law and prosecution can result.

"Fortunately," they add, "nowadays, we don't have any significant problems in this area."

What exactly are the nature and magnitude of these CAP assets? Take the aircraft fleet, for instance. Naturally, since the number of aircraft varies from month to month as some are attrited by accident or disposal and replacements are gained by purchase and through trickle of military types suitable to CAP's mission and surplus to DOD requirements, it is necessary to pick a reference point, in this case the last day of Fiscal Year 1973.

At that time, the Civil Air Patrol's eight regions and 52 wings operated a total of 780 aircraft ranging from 65-horsepower Aeronca Champions, Piper Cubs and Cessna 120s to the husky deHaviland U6A Beaver (better remembered as the L-20) and twin-engine Cessna 310s and Aero Commanders. Some 450 of these were former military aircraft given to CAP by DOD including types like the Piper L-4, Stinson L-5, Aeronca L-16, Cessna O-1 (formerly known as the L-19), Beechcraft T-34 Mentors and Beech C-45s. The remainder of the fleet is made up of civilian aircraft either purchased by CAP or acquired by donation from private individuals. These include virtually every single-engine Cessna from 120s to 185s; six models of Piper aircraft; civilian Stinsons and Aeroncas; Navions, Bellancas, Mooneys, Ercoupes, Luscombes and Taylorcrafts. Acquisition value of the fleet at that time—$8 million. Since then the 38 Cessna T-41s have been added and Helio Couriers are becoming available through military sources.

Air Force personnel who staff the Logistics office at National Headquarters point with some satisfaction at the way the aircraft replacement program is working. In 1972, as an example, CAP disposed of 118 aircraft which were no longer capable of being effectively used in the program—primarily due to their age and the prohibitive cost of restoration—and during the same period acquired 52 by donation and 51 by purchase for a net loss of only 15 aircraft. Aircraft are disposed of only with the approval of National Headquarters and then only by closed bid procedures. Funds derived from such disposal initially go into a special fund maintained at National Headquarters and are subsequently disbursed back to the wings and units on an equitable basis for use exclusively in acquisition of new aircraft or maintenance and upgrading the existing inventory. Despite the skyrocketing costs involved in acquiring aircraft as well as in their maintenance and operation, Civil Air Patrol has managed to keep its fleet at or about the 800 mark for most of the past 20 years.

Aircraft and other property acquired from the military services goes into the CAP inventory at a figure roughly 20 percent of its original DOD value. Within the context of DOD property disposal regulations, this means that the property is going to be utilized in such a way as to indirectly continue its usefulness to the government (in performing those missions authorized in the Public Law and in Air Force directives) and therefore the losing organization can claim a "cost recovery" in its disposal to CAP. By rule of thumb, this cost recovery amounts to approximately 20 percent of its original acquisition cost. Thus, the equipment is given this value when entered upon CAP property accounts. How much such property comes CAP's way in any one year is directly proportionate to several factors—(1) current mission requirements of the active military services, (2) requirements of the CAP mission, (3) Air Force (CAP-USAF) evaluation of that requirement.

The manner in which military mission requirements affect the flow of such property (there must be a continual flow since the property is "used" to begin with and its further useful life ordinarily is quite short when CAP acquires it) can be observed in the statistics for Fiscal Years 1965 through 1972.

Year	DOD Property Acquired (CAP acquisition value 20% of original DOD cost)
1965	$1,760,000
1966	1,002,000
1967	737,000
1968	982,000
1969	1,419,000
1970	2,629,000
1971	3,740,000
1972	3,070,000

(Totals are for all categories authorized including aircraft)

As DOD commitments in Viet Nam reached their peak, surplus property available to CAP to perform its mission was drastically curtailed. When Southeast Asia commitments began to wind down the amount of surplus equipment increased. Air Force authorities point out, however, that while this figure hit a high in the 1972/1973 period "a drastic reduction" in DOD property was expected to begin in 1974 and extend through the mid-70s.

Another measure of Federal interest in the Civil Air Patrol is the amount of Air Force "appropriated" funds which go toward indirect and direct support for its auxiliary. Fiscal Year 1973 figures looked like this:

Category	Amount
Military Personnel (At standard rates)	$3,805.900
Civilian Personnel Compensation	1,344,900
Travel	604,900
Transportation of Things	4,300
Rents & Utilities	26,000
Communications	56,900
Contract Maintenance	100
Purchased Services	182,000
Supplies	16,800
Ground POL (Petroleum, Oil, Lubricants)	47,800
Equipment	14,400
Aviation POL (for AF aircraft)	59,500
Aviation POL (for CAP aircraft)	201,500
TOTAL	$6,365,000

Of these, the following amounts constitute direct support by the Air Force:

a. Communications: FY73—$31,524. This expenditure was made to reimburse CAP members for telephone calls made while participating in USAF authorized search and rescue missions and test exercises, civil defense test exercises and disaster relief activities.

b. Ground POL: FY73—$47,201. This expenditure represents the total amount paid to CAP members for the automotive gasoline and oil they used while participating in USAF-authorized search and rescue missions and test exercises, civil defense test exercises and disaster relief activities.

c. Aviation POL: FY73—$201,463. This expense is reimbursements made to CAP members for the aviation fuel and lubricants they used during USAF-authorized search and rescue and disaster relief missions.

One of CAP's nagging problems, particularly at the unit level, is that of adequate quarters providing not only space for administrative functions, but also for training, storage, communications and operations. Obviously, cadet and composite squadrons usually have a requirement for a larger area because of the cadet training mission. SAR units ordinarily can get by with quarters that provide an administrative office and an operations/communications room. A group headquarters usually can be accommodated in modest quarters since it is largely an administrative element of command. Wing headquarters require relatively large amounts of office space since the wing is the major element of management and is the "residence" of the wing's corporate officer—the commander. For a civilian volunteer, the legal responsibility of the wing commander is awesome and to perform his job effectively a large staff of specialists is required.

Civil Air Patrol organizations across the country call a wide variety of facilities "home." Church basements, school all-purpose rooms, city/county/state/Federal buildings and reservations where there is surplus space available, military installations where space exists excess to the requirements of the service tenants, airport hangars and pilots' lounges. In many areas, CAP units have acquired World War II-type quonset or pre-fab buildings, moved them to land made available by local communities or businessmen and refurbished them for use. As in the case of aircraft and vehicles, this requires hard dollars. In some instances, the costs can be defrayed from the unit treasury. In others, the members combine their individual donations to get the job done. Some enterprising commanders get local builders and suppliers to donate the time and materials.

Wherever possible, a concerted effort is made to house Civil Air Patrol units on military installations. Facilities in these installations are authorized providing there is space available and providing "CAP occupancy does not interfere with the mission of the DOD agency." CAP forces are more "at home" in military surroundings since they are uniformed, disciplined and the USAF auxiliary. Further, such facilities usually are better suited to CAP activities. Often active duty military personnel and DOD civilians turn out to assist the unit in a variety of ways.

In 1973, the Civil Air Patrol was being provided space on more than

100 DOD installations and the availability was "on the increase." It is one of the responsibilities of the AF-CAP liaison officers assigned to the regions and wings to coordinate with DOD facilities in their areas of responsibilities; determine if space is available and make the necessary arrangements with the responsible agency for CAP occupancy. Such occupancy is provided for by a formal instrument (permit/license) executed between the wing commander in that state and the authorized representative of the DOD agency. Such a license can be terminated at any time by either party. Most of them are written for a period of five years and are renewable providing the active establishment does not need the space.

Normally, the host service/base provides the necessary sanitary facilities, water, light and heat at no cost to CAP. No Air Force funds are advanced for refurbishing, rehabilitating, improving or maintaining the facilities. However, it is not uncommon to find active duty personnel working alongside CAP senior members and cadets, volunteering their time, expertise and often money, to put these facilities in shape. In some cases, National Guard quarters for instance, CAP and the host agency find joint use agreeable. The service uses the facilities during the normal work day and CAP uses them nights and on weekends when no drill or duty periods are scheduled.

Several CAP logistic problems will be eliminated if the Congress acts favorably on legislation placed before it in 1974. Generally referred to as the "new CAP supply bill," it will amend the U.S. Code to permit the Air Force to increase its support in certain critical areas. The new legislation—which has been approved both by the Air Force and DOD and sent to Congress for action—would increase direct support to CAP in three major areas, (1) provide a modest level of reimbursement to CAP members for their out-of-pocket expenses incurred while away from home on Air Force-ordered missions, (2) provide a similarly modest level of reimbursement for the "variable flying costs" incurred on such missions, and (3) provide for "free" uniforms for certain Civil Air Patrol cadets who meet strict criteria.

CAP members are often called upon to participate in missions so far distant from their homes as to make it impractical for them to return home for food and lodging. Though no exact data have ever been compiled on the costs incurred by members in such situations, it is possible to estimate these costs by examining a typical mission and making certain reasonable assumptions.

A typical search and rescue mission lasts about three days. Thirty personnel, flying 12 aircraft, will be used throughout. While most participants will find it possible to commute back and forth to their homes during the mission, approximately 20 percent or six individuals, will not. This means that on a typical mission, the cost of 18 days worth of food and lodging must be paid for by individual members, not to speak of the travel expenses involved in getting to and from the mission site and oftentimes even loss of wages.

The provision in the proposed amendment to Section 9441 of Title

10, United States Code, which calls for payment of per diem for travel, food and lodging to CAP members while participating in authorized Air Force missions away from home, is designed to correct this inequity.

General Westberg, National Commander, puts it this way:

"It appears to be something less than fair for this nation to expect individuals, who are performing a humanitarian service for the country at considerable personal sacrifice and expense to themselves, to also pay for their food, travel and lodging while away from home."

Another provision of the proposed amendment would provide CAP members reimbursement for the variable costs they incur while flying on officially-authorized missions. CAP members are now reimbursed only for the fuel and lubricants they use while flying on such missions. Under this provision, they would also be paid on a pro rata basis for other variable cost items, such as engine, radio, airframe, and instrument repair, which are just as much a part of the cost of flying missions as are fuel and lubricants.

It is a fairly simple matter to establish costs for POL. It is not quite so easy to figure other variable cost factors. However, a reliable formula for doing so has been established through experience over the years. This formula is expressed thusly: rated horsepower of the engine x .06-total variable costs. One-half of this total is fuel. The other half is for the repairs it takes to keep the equipment operational.

According to General Westberg, "It costs CAP approximately $7.23 in aviation fuel and lubricants per hour flown, and it is this flying cost for which CAP members are currently being reimbursed. However, since this is only half the amount of the total variable costs for flying that are actually incurred, the other half must be absorbed by the individual CAP members.

"It is to correct this inequity that this provision has been inserted into the proposed amendment. Its approval would permit CAP members to collect a higher percentage ($14.46 average per hour) of the total variable flying expenses they incur in flying on authorized missions.

"In terms of additional cost to the government, this provision would not be expensive. Since CAP is now flying some 27,000 hours per year on authorized search and rescue missions, this would amount to a total cost of about $391,000 annually, as compared to the total current annual reimbursement of $201,500."

Another of the provisions of the proposed amendments would provide free uniforms to certain groups of CAP cadets. Under one option being studied, each "newly enrolled cadet" might be supplied with a fatigue uniform free of charge. Then, upon completion of Achievement I under the Cadet Training Program, he could receive a Class A uniform, also free of charge. The annual cost of such a free clothing-issue program has been calculated this way:

The male fatigue uniform, consisting of shirt, trousers, and cap, costs $6.79. The female fatigue uniform, consisting of shirt and pants, costs $11.27. Assuming an average annual input of 16,000 new cadets, composed of 12,800 male and 3,200 female cadets, the expenditures on fatigue uniforms could be anticipated to be $122,976.

Items of the Class A uniform that would be issued to each cadet completing Achievement I would cost $34.35 for male cadets and $60.97 for females.

Assuming that 8,000 cadets, 6,400 male and 1,600 female, complete Achievement I each year, the annual cost of issuing Class A uniforms would be $317,392.

The annual costs for both types of uniforms results in a total annual cost of $440,368 for the entire free-uniform issue program. There is no plan to provide free uniforms to members already in the cadet program, inasmuch as the majority in this category would already have purchased them or obtained them through excess property channels.

General Westberg believes that "the free-uniform program as visualized for cadets would serve as a powerful incentive to those youngsters who would like to join CAP, but who hesitate to do so because of the personal costs involved". Belonging to CAP, especially for cadets who are still dependent on their parents for most of their financial requirements, is a rather expensive business, as this itemized estimate of annual costs for an average cadet in his second year of membership reveals.

Description of Cost	Amount
National dues	$5.00
Wg and/or Sq dues	6.00
Completion of 3 contracts @ $2 ea	6.00
1st class postage for 3 contracts @ $.50 ea	1.50
While attending Type A Encampment	
Meals	15.00
Personal expenses (laundry, haircuts, etc)	20.00
Squadron activities (overnight trips, tours, meals, etc.)	30.00
Miscellaneous (cleaning of uniforms, extra ribbons, patches, etc.)	10.50
Total annual cost	$94.00

NOTE: This estimate does not include uniform costs, costs for participating in any nationally-sponsored special activity, or flight training.

"Furnishing free uniforms," Westberg concludes, "to new cadets and those second-year cadets who are demonstrating a sincere and continuing interest in CAP would go far toward helping them defray the costs of their membership, at least during the first and the most critical years of their affiliation with CAP."

The proposed legislation would have one other salutory effect on CAP. It would permit the Secretary of the Air Force (with the approval of DOD) to "arrange for the assistance of other military departments, Federal departments or agencies." The current law appears to limit the secretary's authority to that of permitting the use of services and facilities of the Air Force or arranging for use of other

DOD facilities. In proposing the legislation, it was pointed out that "the noncombatant services of the Civil Air Patrol are national in scope" and that the assistance of other agencies "would greatly enhance the response and effectiveness of the CAP as an emergency services agency." While the cost of the legislation cannot be definitely ascertained since CAP cannot forecast cadet membership turnover and frequency of national or local emergencies, on the basis of cost data projected over a five-year period an annual average estimate of $1,017,200 has been established and General Westberg emphatically points out:

"Even conceding the possibility that the entire package can result in an additional $1 million annual cost ($1 million over that presently appropriated under the original statutory authorization), this would still amount to only approximately 10 percent of the total value of the vital services performed by CAP for this nation every year."

To a significant extent, CAP has, over the years, made itself self-sustaining in many areas. At the time CAP published its 1974 Report to Congress, it reported its financial status and activities for Fiscal Year 1972 and 1973 like this:

INCOME	For the year ended June 30.	
	1973	1972
Members dues and charter fees	$517,970	$566,431
Members contributions	33,795	36,209
Interest earned	22,144	16,601
Sale of educational material	279,679	242,964
Amarillo Depot receipts	64,180	33,655
Other	5,761	5,175
	923,529	901,035

EXPENSES	1973	1972
Cadet activities	$162,546	$167,841
Senior activities	10,076	14,479
Subscriptions	34,100	34,852
Public relations	6,814	3,406
Insurance	83,265	86,124
Machine rental	46,672	54,652
Corporate employees	25,535	22,337
Regional and national chairman fund	15,216	10,554
Administrative support	20,229	11,215
Equipment maintenance and expense	16,658	13,298
Contingency reserve	14,104	4,215
Art and art supplies	4,104	2,673
Awards	6,065	4,121
Other equipment operations and maintenance	3,255	2,852
Administrative	12,409	10,675
Protocol	566	633
Professional salaries and expense	23,782	33,748
General aviation		27,011
Self-insurance expense	1,489	7,220

National scholarship fund	22,144	16,601
Budget items, 1972 and 1971 respective	27,668	24,578
Publication production expense	239,473	208,129
Depreciation	7,708	6,703
Business members expense	463	512
Amarillo Depot expenses	65,064	23,049
	849,405	791,478
Excess of income over expenses	$74,124	$109,557

It is evident that CAP's major source of income is the dues of its members. In other words, Civil Air Patrol members not only risk their lives, give of their time and donate significant amounts of their own financial resources, they pay for the privilege of doing so.

Two items listed under "income" in the financial statement warrant amplification since they represent two ways in which the Civil Air Patrol is becoming more self-sufficient. These are the "sale of educational material" and the operation of the "Amarillo Depot."

In the mid-1950s, CAP National Headquarters realized that if it wanted to provide aerospace education materials to its cadets and to schools interested in extending such courses to their students, it was apparent that the Civil Air Patrol would have to take the bone in its teeth, as it were, and develop its own capability. Not only were materials then available from outside sources inadequate in terms of applicability to CAP's objectives and purposes, they were in scant supply and would represent a prohibitive expense. Under the guidance of the full-time, professional educators employed by the Air Force in behalf of CAP, the corporation developed its own educational materials, publishing and distributing them on a pay-as-you-go basis. Pricing was established so as to maintain a revolving fund from which the cost of updating and/or developing new materials could be defrayed. That program ultimately developed into the "CAP Book Store", an independent operation of the corporation which now provides, at reasonable cost to members and units, not only aerospace education materials, but also insignia, decorations, uniform items, training pamphlets, regulations and manuals. The operation pays for itself and even occasionally results in a small profit.

The other operation, the so-called Amarillo Depot, came about because of the huge costs involved in modifying surplus military aircraft acquired by CAP to meet the criteria for an FAA Airworthiness Certificate and for conducting overhaul of these aircraft when required.

The depot, at Amarillo, Tex., is managed by Fred Chesser, a retired Air Force supply specialist now employed by the corporation. It stocks the hard-to-come-by parts for as many of CAP's military-type aircraft as possible making them available to CAP units at a few cents above cost. Those few cents go to defray the cost of operating the depot. An Amarillo CAP squadron commanded by Lt. Col. Earl Parks (who in civilian life operates an aircraft service facility) donates its time in assisting Chesser in operating the depot.

The depot also represents a "central buying center" for purchasing items like ELTs, VHF-DF sets, tires, batteries, spark plugs, instruments, etc. This enables CAP to benefit from wholesale prices and bulk discounts. Again these articles are sold to units and members for their aircraft at prices which include only cost plus the amount needed to defray depot overhead. The depot also handles the responsibility of contracting with outside vendors for major overhaul programs like the one conducted to bring the 38 Air Force T-41s into the active inventory. CAP units can arrange with Parks for overhaul of individual CAP aircraft. Parks and the members of his unit who do the work charge only the "bare cost" of the work plus the spare parts obtained from the depot at depot costs. The result is a significant saving for Civil Air Patrol units.

One other service CAP headquarters is providing for its units is a system whereby wings can finance aircraft purchases at low interest rates. A wing wanting to purchase an aircraft and satisfying its regional commander of its capability to support the aircraft from both a financial and utilization aspect, can obtain the aircraft at five percent "add-on" interest with the corporation guaranteeing the purchase to the bank.

Individuals new to the Civil Air Patrol as well as most outsiders take for granted the term "non-profit organization" used in the Congressional charter is little more than a figure of speech. Certainly, they reason, it is not a corporation in the sense of the General Electrics, the Fords, the Standard Oils and the Monsanto Chemicals.

The term non-profit also tends to belittle the magnitude of the job involved.

"Nothing could be further from the truth," Bob Walker observes, "CAP is big business and in addition to all the problems of a profit-making venture, it has some hairy ones of its own."

Cessna O-1 Birddog (better known to Korean vintage troops as the L-19) today is one of the mainstays of the corporate CAP aircraft fleet. O-1 with its outstanding visibility, good power-to-weight ratio and short-field performance is especially valuable in air search.

CHAPTER 15

Where Do We Go From Here?

For nearly 35 years the Civil Air Patrol has been an important part of the American aviation/space scene: performing operational missions ranging from wartime anti-submarine patrol to peacetime search and rescue; helping to develop the aviation/space leaders of tomorrow, if you will, latter day Goddards, Bormans, Rickenbackers, LeMays, Cochranes, Earharts and Lindberghs; and working through the educational system toward a greater public awareness of the importance of aerospace in our society.

Getting the job done—in this case meeting it's own self-established objectives—has not always been easy for CAP. Like any large organization it has suffered from its growing pains and internal conflicts. It has gone through periods when rapport with its parent organization, the U.S. Air Force, has waned with the result that badly needed support appeared to be withering. It has struggled to maintain its integrity and capability during periods of post-war apathy when large segments of the public move dangerously in the direction of isolationism and in times of broad public disenchantment with politics, politicians and their government.

In a way, the Civil Air Patrol has reacted in these times very much like the microcosm of America it is. When the going gets tough, the men and women of CAP put aside their individual differences, fall in shoulder to shoulder and with a fierce determination, move to meet the common threat head-on. This capability will stand them in good stead in the years to come for, as in the past, the future doesn't appear to promise any bed of roses for CAP.

A paradox is emerging on the Air Force side. On the one hand maintenance of a strong, effective Civil Air Patrol is going to be more important than ever. The USAF now faces strong competition for the high caliber young men and women it needs from the Army and the Navy. Without a draft, the services must maintain their strength by attracting volunteers. For many years, much of the manpower requirements for the Army and to a slightly lesser degree, the Navy were satisfied primarily from the draft. This tended to make available to the Air Force and the Marine Corps more of the highly motivated young men and women—those who decided upon a military career on their own. By and large, these young people also represent a high level of intellectual and physical ability. To a great extent this type of young man and woman will continue to be attracted to the USAF and the Corps, but already both the Army and Navy are gearing up to make career opportunities in those services appear just as attractive. Competition for the motivated ones is going to become keen, even bitter. In such an environment, the obvious value of the Civil Air Patrol cadet program to the Air Force assumes even greater significance.

On the other hand, defense dollars—those dollars needed to insure our national security— are getting more difficult to come by and at the same time the cost of defense continues to be driven up by inflation. Most economists do not see the value of the dollar increasing. Changes are occurring in national priorities. New requirements to spend huge amounts of tax dollars are emerging at the same time the need for additional funding of existing national programs increases. In other words, here also competition is becoming keen and often bitter. Not only is it becoming increasingly difficult for the Department of Defense to get needed dollars, it is becoming more difficult within DOD for the Air Force to obtain funding for the programs it considers necessary to national defense. Competition among the services for available funds always has been fierce. It can be expected to get more aggressive. In this context, the more knowledgeable advocates the Air Force has at the grass roots level of America and the better these citizens and their representatives in the Congress understand the need to maintain aerospace supremacy as an instrument of national security, the better chance the Air Force has to insure that its critical programs are adequately funded.

In this same vein, the primary operational mission of the Civil Air Patrol, search and rescue—assumes even greater importance. The Air Force continues to adjust its operational priorities to keep in line with equipment and manpower cuts. Already CAP is performing up to 80 percent of the actual search and rescue flying hours required for the Air Force (specifically ARRS) to discharge its responsibility under the National SAR Plan. This may be expected to increase until CAP's volunteer aircrews find themselves responsible for almost all of the USAF commitment. The following exercise in statistical brainstorming puts CAP's SAR mission into perspective:

An Air Force second lieutenant, unmarried and a rated pilot or observer on flying status receives (in 1974 figures) $857.68 a month—$600.90 in base pay, $100 in flying pay, $108.90 for quarters allowance, and $47.88 for rations.

The Civil Air Patrol represents a force-in-being of at least 2,000 SAR-rated pilots alone. To maintain a similar force available on a stand-by basis would cost the taxpayer some $1.7 million a month.

The 2,000 Air Force pilots stationed strategically in local communities across the nation in all 50 states, the District of Columbia and Puerto Rico would have to be dedicated to the SAR mission and therefore would not be available for any other duty. They would cost the taxpayer $20.4 million a year.

What about aircraft? The military services no longer maintain an active inventory of the type of aircraft—like the Cessna Birddog, the DeHaviland Beaver and the Helio Courier— required for SAR, particularly for SAR at a "reasonable cost." Let's buy our 2,000 pilots 2,-000 contemporary lightplanes of sufficient power and flexibility and with adequate equipment. At 1974 prices this amounts to an expenditure of $40 million. Add $5 million in spare parts inventory and then amortize the total investment over a 10-year operational life—4.5 million a year.

We still haven't considered operating these aircraft. CAP flew more than 27,000 hours on Air Force-ordered SAR missions in 1973, for instance. On an average, fuel consumption for CAP aircraft can be calculated conservatively at 10 gallons per hour or 270,000 gallons of fuel. With 1974 prices—$.60 a gallon on the average—this amounts to $162,000. Throw in 15,000 quarts of oil at $.85 a quart—$12,750. Maintenance at $2.50 per flying hour adds another $67,500. Stretching a point, we won't allow for any attrition due to accidents, storm damage, what have you. Grand total for operating this, our mythical fleet of SAR craft—$4.7 million a year (including the $4.5 million per year to amortize the initial acquisition cost).

Crew cost, aircraft cost and operational costs now total a whopping $25.1 million a year.

By contrast, in 1973, 27,000-plus hours of SAR cost the Air Force approximately $259,000 including commercial communications primarily used in gathering all the bits and pieces of information needed to intelligently conduct a search.

In round numbers, the Air Force spends about $6.5 million a year in support of the entire CAP program—operational missions, cadet program and aerospace education. This included that 27,000 hours of SAR flying.

Result of our little statistical game—a $21 million saving to the taxpayer. But, in all candor, it must be admitted that our force of 2,000 Air Force SAR crewmen and their 2,000 airplanes could not come close to representing the overall potential of the present CAP force of 18,000 pilots and 6,000 airplanes and the 30-plus years of hard-to-come-by SAR experience of its personnel.

General Westberg, who answers to the Secretary of the Air Force for the Civil Air Patrol and for its readiness (or lack of it) to perform those missions authorized by the Congress, is a firm believer in the capability of CAP to perform the SAR mission and the requirement for it to continue in this role. He cites a 1973 mission and its evaluation by Air Force and FAA officials as indicative of the high degree of expertise in SAR represented by CAP, especially its mission coordinators.

An airliner disappeared on a short 100-mile route segment during a period of inclement weather. CAP, as well as other agencies in the area including aviation elements of other military services, were alerted. Arkansas Wing mission coordinator LTC Charles M. McKinnon got his briefing from the Air Force mission controller at the Central Rescue Coordination Center, Richards-Gabauer AFB, Mo., and went to work. Almost immediately he was challenged by a military commander of a cooperating agency for his decision not to launch search aircraft at night and during a frontal passage in the search area. McKinnon's rationale was simple: If the aircraft had crashed and burned there would have been reports of fire; if there were no fire, search aircraft would have no chance of sighting it in the pitch darkness; and, finally, the line squall represented a real danger to the safety of light aircraft.

McKinnon's position was unacceptable to the other commander who immediately launched helicopters. Within an hour, one of these crashed killing its crew of four. That commander immediately grounded the rest of his aircraft until day break.

Westberg offers these three letters sent to him at CAP National Headquarters as evidence of this CAP mission coordinator's outstanding capability, adding at the same time:

"Thank God, we have a lot more like Bill McKinnon in CAP!"

"Please convey my personal appreciation to Lt. Col. Charles M. McKinnon, his staff, and all participating members of the Arkansas Civil Air Patrol for their outstanding performance during the recent Search and Rescue Mission 43-076 from 28 to 30 September 1973. This mission involved a missing Texas International airliner enroute from Eldorado, Arkansas, to Texarkana, Texas, on 28 September 1973 with eleven persons on board.

"Lt. Col. McKinnon was assigned the difficult task of Search and Rescue Mission Coordinator (SMC) at Magnolia, Arkansas, near the last known position of the aircraft. Lt. Col. McKinnon immediately organized a rescue operations center and carefully analyzed available rescue resources and weather conditions. Within hours after notification, search aircraft from the Civil Air Patrol (CAP) were launched for the initial route searches along the missing aircraft's planned flight path. Concurrently, additional SAR resources from the United States Army, United States Air Force, and Air National Guard were requested to assist in the search.

"Due to marginal weather conditions, the first day of search, negative results were achieved in locating the aircraft. On the second and third days of the mission, both military and CAP forces worked hand-in-hand out of Lt. Col. McKinnon's operations center. An average of over 70 aircraft searched daily, the peak being 95 aircraft on the third day. Control and coordination of a force this size is indeed a monumental task, considering crew briefings, assignment of search areas, administrative logging of sorties and flying hours, and arrangements for the billeting and feeding of search personnel.

"I personally observed Lt. Col. McKinnon's operation on 29 September and found it to be an absolutely outstanding effort in all respects. The orderly and efficient manner in which the operation was organized resulted in exceptionally efficient and controlled utilization of all search forces. Four hundred thirty-one sorties and 1099 flying hours were logged during the three day search—a truly large, significant search effort!

"The close and coordinated civilian-military relationship maintained during this mission was outstanding and professional in every manner. Brigadier General Sullivan, Commander, Aerospace Rescue and Recovery Service, was highly impressed by the CAP personnel at the Magnolia rescue operations center and the overall operation. The expertise and professionalism displayed throughout this mission is highly commendable and a direct reflection of Lt. Col. McKinnon's conscientious devotion to duty in carrying out the Rescue motto, "That Others May Live".

"Please congratulate Lt. Col. McKinnon and his entire staff for their "can do" attitude. Their untiring efforts and the long hours expended on this mission were in the best tradition of the Civil Air Patrol. Well done!"

BILL A. MONTGOMERY, Colonel, USAF
Commander
43rd Aerospace Rescue and Recovery Squadron

"The search and rescue mission for the TIA aircraft in Arkansas is a classic example of joint military/civilian cooperation. Although dampened by the untimely death of the crew and passengers, it was, nonetheless, especially gratifying to note the close teamwork between CAP personnel and the other units involved.

"The effective support and notable contribution provided by Lt. Col. William J. Williams and Lt. Col. Charles M. McKinnon relate directly to their outstanding knowledge and extensive qualifications in rescue operations. The professionalism displayed by CAP personnel in the performance of their assigned responsibilities contributed directly to the success of this mission.

"Please convey my sincere appreciation on behalf of this command to the personnel involved in this mission for a job 'well done.'"

GLENN R. SULLIVAN, Brig. General, USAF
Commander
Aerospace Rescue and Recovery Service

"I would like to bring to your attention the excellent manner in which the search for the Texas International Airlines, Inc., Convair was conducted by the Arkansas Civil Air Patrol under the command of Colonel McKinnon. He and his members of the Civil Air Patrol unselfishly spent many many uncompensated hours and a great deal of their own personal funds in effort to locate the aircraft.

"In addition, I feel that I must comment on the professional manner in which the search was conducted. The orders issued by Colonel McKinnon were clear and precise. It was obvious to the observer that the operation was being conducted by professionals and left no doubt that the aircraft would be located regardless of the difficulties encountered.

"Their actions spoke well of their state, its citizens, the CAP, and your leadership."

Sincerely yours,
E. D. Dreifus
Investigator in Charge
National Transportation Safety Board

Even with its many years of search and rescue experience, Civil Air Patrol mission coordinators and SAR aircrews are facing new challenges as developing technology increases the chance of saving more lives and at the same time brings with it new problems. The emergency locater transmitter (ELT) came into its own with the Armed Forces during the Viet Nam conflict and has now become a part of the general aviation scene.

A small, compact, solid-state, radio transmitter which when triggered by an impact of predetermined G force, begins emitting a warbling signal on 121.5 Megahertz (the international very high frequency emergency channel), the ELT was made a mandatory piece of equipment on most general aviation aircraft in 1974. CAP and other SAR forces have begun recording an increase in the number of "saves", a number directly attributable to the ELT. The ELT, when it operates properly, permits the search forces to home on the signal with very high frequency direction finding equipment. It can lead searchers to a crash scene in a much shorter time thus enhancing the survival ratio of casualties.

ELTs, however, do not always operate properly and, more important, they frequently go off in transit, when an aircraft makes a hard landing, on the shelves of the FBO when bumped or dropped. Within a single three-month period, the FAA reported more than 900 false ELT events. CAP forces around the country found themselves "run ragged" checking out the huge number of reports eventually tracking most of them to airport hangars and shops or parked planes. But, every report must be checked out. Neither the FAA, the Air Force nor CAP can run the risk of the ELT signal that isn't checked out being the real thing. This is placing a heavy, new workload on CAP SAR forces.

Still another ramification of the ELT is the need for airborne direction finding equipment. It costs upwards of $360 per aircraft to equip a search plane with first class equipment— a price tag of more than a quarter million dollars to outfit just the corporate fleet. Yet that is exactly what CAP is doing with local, wing and national funds as fast as the money can be made available. Still a full capability to perform ELT search won't be in hand until the entire fleet also is equipped to conduct night search and search under instrument conditions and until there are a sufficient number of aircrews also trained and licensed (FAA instrument certificated) to man available corporate and member-owned aircraft. The Civil Air Patrol, like the military services, is finding it costly to keep up with the new capabilities provided by the steadily advancing technology.

While General Westberg is proud of CAP's SAR performance and confident of its ability to develop an even higher degree of proficiency, he doesn't relegate either CAP's cadet program or its aviation education responsibility to the back burner. He considers both of them as, and in some respects, more important than the operational mission. Westberg also points to another plus represented by the CAP.

"Civil Air Patrol," he explains, "provides an Air Force image—and a good one—in nearly 2,000 communities throughout the nation where otherwise there would be none. This constant, day-to-day, grass-roots presence is of incalculable value."

"There also is a real benefit accruing from the aerospace education program," he continues, "but it is hard to measure. I defy anyone to put a yardstick on and say just how much progress is made in any one year. We do know, however, that we are making progress. On an average, there are 185 aerospace workshops held each year either under CAP sponsorship or staged jointly by other aviation organizations and universities with CAP cooperation. More than 10,000 teachers are exposed to aerospace education techniques and methods each year through these workshops. These are people we want to get the word about this aerospace business. They, in turn, influence the young people. In a few years the snowball effect will enrich the country in aerospace-minded citizens. If you want a good illustration of how we and the entire aerospace industry failed to do our job before, just look at the SST. The vote on that indicates that aviation and aerospace hand't done a very good job. Today aerospace education courses are being taught in some 1,500 high schools in this country. All or part of the Civil Air Patrol aerospace education curriculum is used in the majority of these schools. This is an important step in the right direction."

Westberg also gives the CAP cadet program high marks as a means of insuring that future citizens will better understand the importance of aviation and space and as a springboard to launch more young people toward careers in these industries.

"CAP is making a real contribution in this line," he asserts, "preparing young people for possible careers in aviation or the aerospace industries or, for that matter, any of the supporting industries. It also motivates them toward careers in the military services, not necessarily the Air Force, but any of the services. Their training in management and leadership can be put to use in any industry or profession. They are learning self discipline and developing traits of leadership that will be most valuable later on in life.

"The cadet program also is turning out young citizens who not only are knowledgeable in the areas of aviation and space, but also tend to be articulate on these subjects. One of the areas in which we need the most help—and by we I mean not only CAP and the Air Force, but the entire aviation/space industry—is articulate people able to tell-it-like-it-is before public forums. The public can and will help if they understand. We have to help them understand exactly what their stake in aviation and space really is."

Westberg indicates that all too few CAP senior members fully grasp the seriousness of this problem and do something about it. Even among active duty Air Force personnel, he says, there is a great dearth of spokesmen who fully understand and can articulately explain just what the needs of national security are and exactly what the Air Force role is in satisfying these needs.

With regard to the capability of CAP members to articulate the CAP story, Westberg also voices concern that a lack of "complete understanding of the Civil Air Patrol, its objectives and purposes and its missions," on the part of senior members is having a serious effect on both recruiting and retention.

"If you don't know your own organization," he declares, "how do you expect to sell it and keep it sold?"

The National Commander's concern in this vital area appears justified. All too many CAP senior members, in many cases the more dedicated and active, do not take the time to become knowledgeable in terms of the complete program. Certainly the SAR people know their business. The communicators know their business. The ground rescue people know their business. The managers and administrators know their business. In their individual narrow areas of responsibility, they are experts. To an alarming extent, however, they cannot "sell" the total program because their overall grasp of the big picture is sketchy at best. Westberg points to the level I Senior Training Program as a means of insuring that all members acquire a better grasp of exactly what CAP is, how it came into being, what it does and how it does it. But, he expresses a deep concern that not enough senior members are completing the course, particularly as he puts it, "the old-timers who think they don't need it."

The need for CAP members to be able to clearly and concisely discuss their organization with representatives of the public is closely aligned with the necessity to make CAP better known to the media. By and large, CAP members, including all too many unit commanders, do not understand the necessity for an aggressive public information program. When they do a good job, successfully complete a mission or reach a milestone they tend to take the position that the media readily understands the importance of that mission or the significance of that milestone and thus should come "beating on their door for information." They fail to realize what seems important to them may or may not seem important to the local newspaper editor or TV station news director. They do not grasp the very real fact that in many instances the decision to run a particular story or news clips depends entirely on when the information was made available and even more important how it was packaged. In other words, how the information was presented and whether or not it was placed in perspective with relation to the achievement of larger goals and objectives which clearly are of public interest.

CAP National Headquarters is working to correct this situation. A professionally-constructed public information training program is being carried out and, with the unequivocal endorsement of General Westberg, the critical importance of the information function is being emphasized to commanders at all levels.

In Westberg's opinion the importance of the information program "cannot be over emphasized."

"It isn't enough to accomplish the mission," he declares, "we must tell the public about it. The Civil Air Patrol is almost entirely dependent upon the public for support either directly in terms of helping local units fund their program and in terms of state appropriations (where they are available) or with regard to the indirect but indispensable support provided by the Air Force with approval of the Congress. The only way the Civil Air Patrol can continue to enjoy this support is (first) accomplish the task successfully; complete the mission, and (second) make sure that adequate, timely, factual information is made available to the media."

The essence of CAP's future course as it passes the midpoint in its fourth decade Westberg sums up in just four words—"do more with less."

"We are entering a lean period," he warns, "with the energy crisis causing a cutback in the amount of fuel available. Efficiency of operation is a must. Aircraft scheduling must be closely monitored, both as to purpose of mission and pilot qualifications. Currency and pilot proficiency must be reviewed in relation to type aircraft and weather conditions.

"Overall support from the Air Force is, of necessity, going to be available at a lesser level. Civil Air Patrol-USAF—the Air Force side of the house—has to compete for funding along with all the other missions the USAF must accomplish, just as the USAF must compete DOD-wide. Not only can we expect funding cutbacks, we can and should expect manpower cutbacks. This also applies to Air Force airlift available to CAP personnel and to the aircraft rental program of the USAF—CAP liaison officers. This will directly affect the capability of the liaison officer to support his individual wing or region. There is some indication that these cutbacks also may affect the Air Force Reserve Assistance Program as well as the overall level of encampment activity at Air Force bases."

There are some steps the general recommends be taken at the local and state level which can serve to attenuate the effect of a diminishing level of USAF support.

"Get out and get those state appropriations," he urges, "CAP programs to develop a public awareness of aviation/space; its programs to help develop stalwart, capable future leaders; and its proven capability to assist the public in times of natural or man-made disaster should be of even greater importance to the states than they are on the Federal level."

An important step in this direction, he feels, is "improving CAP's relationships with state civil defense and emergency services offices, state boards of aeronautics and, of prime importance, local and county law enforcement agencies." The latter he sees as a "difficult area since the old-timers (sheriffs, etc.) are wary of losing any of their traditional prerogatives."

"What CAP has to do," Westberg says, "is to not only convince them, but prove to them they aren't losing a single thing, they are gaining new resources with which to fulfill their commitment to the citizens of their community or county."

Another self-help step, he calls for, is improvement of CAP's flying safety record. It has improved substantially in recent years, he indicates, particularly with relation to the increasing number of hours flown by the Civil Air Patrol on an annual basis. Total flying hours on all activities—actual SAR, SAR and CD training/evaluation, logistic support, individual proficiency, cadet flight training and orientation—was over the 100,000-hour mark in 1973. There still is considerable room for additional improvement, however, and such improvement is of two-fold value. It does, of course, preserve CAP lives and property. It also can serve to make CAP even more acceptable to the total general aviation community, especially to the backbone of general aviation—the fixed base operator.

Almost as an aside, General Westberg puts this forth in conclusion:

"It is worth giving some thought to seeking quality instead of numbers in the recruiting effort."

Although Westberg does not draw the analogy, his thinking here is closely aligned with the current recruiting philosophy of the U.S. Marines—"we need a *few* good men."

Westberg is firm in his conviction that CAP will safely navigate the lean years ahead just as it has the lean years in the past. He also is convinced that "in the long run, the Civil Air Patrol will be all the stronger for it."

Those convictions are shared by the man who took over CAP's corporate reins late in 1973 and will have that huge responsibility for sometime to come.

Pat Patterson—more formally referred to as Brig. Gen. William M. Patterson, CAP—is no johnny-come-lately to the Civil Air Patrol. He's seen it all. There isn't any facet of the organization, its people or its mission with which he is not intimately familiar.

Patterson has quite a CAP track record. He first became affiliated with Civil Air Patrol— the Maryland Wing—because as an airplane owner/pilot he wanted to "do more than just bore holes in the sky." He started out as a search pilot and soon became the Operations officer for his unit. This was purely a flying squadron. It had no CAP-owned aircraft (or as was the system in those days, surplus Air Force liaison planes on loan to CAP). His natural knack for organization soon got things humming and before long the unit boasted some 30 pilot/owners and was building a reputation in the middle eastern states for its SAR capability.

"We did a job and had a lot of fun," Patterson recalls, "and the unit had a great esprit de corps. Then I found out about groups and wings. I made a visit to wing headquarters. They learned about my intense interest in CAP and the attention I was geared to give it. Before long, they talked me into coming to wing as deputy for Operations. The next step was wing executive officer."

Meanwhile, Patterson had been instrumental in forming a new

squadron in nearby Parkville. At wing he had been exposed to the cadet program and became totally dedicated to its purposes and objectives while retaining his interest in the operational mission. It wasn't long before Patterson, who by vocation is a salesman, sold CAP to the local doctor, minister, a couple of lawyers and a lot of other people. The Parkville Squadron started out in one-room quarters at the Maryland School of the Blind, but there was nothing blind about either Pat Patterson or the outstanding people he recruited to assist him. Within two years, the Parkville Squadron was the largest unit in the Maryland Wing with a membership of more than 350—half the total membership of the entire wing.

This occurred during the command of General Vic Beau and, as Patterson tells it:

"Vic Beau heard about a big squadron in nearby Maryland that was buying all its own equipment, wasn't asking for a thing from National Headquarters and had several thousand dollars in the bank so he came up personally to see for himself."

If there was a secret to the success at Parkville, it was in the action-oriented philosophy set forth by Patterson.

"We did things, we did a lot of things for the community as well as a lot of things in the community," he says.

About this time, the Maryland Wing found itself in trouble. It had no home for the wing headquarters. It's operational capability got so bad it was restricted from participating in SAR activities. On the basis of his resounding success in Parkville, Patterson was tapped as the new wing commander. Things began to hum again. Within a very few weeks, the Maryland Wing had a spacious new headquarters in a hangar at Baltimore's Friendship International Airport. The USAF—CAP liaison officer was provided with quarters adjacent to the wing headquarters. The revitalization of the Maryland Wing was off and running. Assignments as commander of CAP's Middle East Region and vice chairman of the National Board ultimately followed with final elevation to the position as Civil Air Patrol ranking member at the time the organization was contemplating its thirty-fourth year.

Patterson is an independent businessman. As a manufacturer's representative handling six to eight major manufacturers of heavy construction equipment and an equal number of smaller companies, he covers a territory extending from New York to Virginia along the eastern seaboard. While he enjoys the freedom to establish his own priorities and manage his own time, it is a demanding occupation and would appear to pose a potential conflict with the demands of his new Civil Air Patrol assignment.

Ask Patterson how he sustains an active business career and at the same time finds time to discharge his CAP commitments. He smiles and responds:

"There's an old cliche, 'if you want to get something done, go to a businessman.' That's the answer, at squadron, group, wing, region and at the national level. Get some top pros in business and the professions to help out. The only way to go when you do that is up!

"I guess what I am known most for is being a motivator; that and a salesman. I sell CAP to other people and then motivate them to do the job and pull together. I don't know what you would call it, a talent or a gift. But, I seem to be able to get people to become dedicated to a cause and then, working together, get the job done."

Patterson feels that the magnitude of the job as chairman is relative.

"When you are a squadron commander," he explains, "you are inclined to wonder how on earth a wing commander handles all the problems. As a wing commander you look at the regional commander—he not only has the problems of one wing, he has several to worry about. When you are a regional commander you can't comprehend how anyone or even a half dozen men can possibly face the problems of 52 wings. The job of vice chairman seems toughest. After all, he's number two so he has to try harder. It's always tougher to be number two. Then finally, you find yourself in the top slot and realize it wasn't really that tough at all anywhere along the line. In fact, it was a lot of fun. That's the real secret. Keep it fun. If you take the fun out of things it becomes terribly frustrating. Like marriage, if you take the love out of it, it becomes a drag."

The national chairman sees the challenge before him and the other leaders of CAP as "tremendous" adding that he has dedicated himself to doing something about the problems facing CAP in his time frame.

"We've been going through quite a transition," he says, "particularly during the past few years when it hasn't been the best thing in the world to be in uniform. Yet CAP people have stuck together and continued to perform their mission. I don't think they have had a fair shake. We at the national level have kind of let the people in the field down. We haven't given them the guidance they needed particularly at the lowest level— the squadron commanders.

"I don't forget the squadron because that's where the action is. That also is where the fun is; you had better believe it. Up here I am just over head. It is at the unit level where the production is—where they are laying out the missions, getting the young people motivated, getting them through this aerospace education program, building America— building Americans. At this level we have the tremendous responsibility of providing for them a better way of doing all this."

Patterson sees a need for some "surgical attention rather than just a band-aid" and has established several short range objectives to accomplish the surgery. One, he says, is the necessity to build better squadron commanders.

"If I went out to build a profit-making organization," he declares, "with untrained managers at the local level, I'd go broke. Yet we do this with CAP. When we form a unit somewhere in suburbia the guy who can get to the meeting place every week gets named squadron commander. We give him all the regulations and say, 'go to it, go out and build a squadron.' All he gets from then on is a lot of abusive treatment, a lot of personal regrets, a lot of policy thrown at him—dos and don'ts—and before we realize it we have lost him. A lot of good people get turned off from CAP that way. As a matter of fact, we turn away a lot of potential achievers in the cadet program the same way."

Specifically, Patterson plans to work "very hard" at building a good platform of management at the squadron level. He sees the necessity of "reaching right down and having some impact at the grass roots level where the attention is needed."

"Everywhere I go across the country," Patterson observes, "each squadron is run differently. Their commanders each have different philosophies and a different picture of what CAP is all about. One of the things we must have is a standardized squadron commander in this program. We already have a major effort under way to achieve this goal in the new Squadron Commander's Handbook and the pocket-sized Squadron Commander's Guide."

The second major immediate goal of CAP's newest chairman is to improve Civil Air Patrol's flying image.

"We're involved in general aviation, deeply involved," he points out, "yet we have never really got together with the other people. We must get closer to the fixed base operators. We must establish better rapport with the Federal Aviation Administration. We have to get closer to the flying public. To accomplish this, we must improve our flying safety record and become increasingly responsible both in our individual and organizational flight operations."

Abraham Lincoln, Patterson declares, "learned by the methodology we in CAP still are using today. You get a book, you study it, you work with it, you take a test. That's the way they did it 100 years ago. We must improve our way of putting the aerospace education course to these young people. Today successful businessmen use audio visuals. It's faster, more economical, more palatable and also you have a higher retention. A major and immediate goal will be to get with modern methods of education in the Civil Air Patrol cadet program."

Conceding that General Westberg's analysis of tough-to-come-by dollars and a lesser level of Air Force support is on target, Patterson still sees an "outstanding continuing relationship with the Air Force" and is confident CAP will get fair treatment in the budget huddles.

"We're doing a bigger job today than we ever did before for the Air Force," he concludes, "and they know it!"

It would be a major error to conclude that Leslie Westberg and Pat Patterson are alone going to sove the challenges that face the Civil Air Patrol in the 1970s any more than Vic Beau and Harold Byrd alone solved those it faced in the 1950s.

They'll need a lot of help and they'll get it. They will have the help of not only thousands of capable veteran CAP members but also the equally dedicated newcomers to the program—those who will carry on from the cadet program as well as those who, from the sidelines, will look and say "there is the place for me."

They will have the help of those in the Air Force, the Federal

Government, the Congress of the United States and the American public who place full value on the many contributions that the Civil Air Patrol has made and will continue to make to the nation.

And, they will have the benchmarks of more than three decades of proud accomplishment by which to chart their course.

REGIONAL COMMANDERS (as of June 1, 1974)

REGION	COMMANDER
Northeast	Col. Julius Goldman, CAP
Middle East	Col. Jonathan H. Hill, CAP
Great Lakes	Col. Robert H. Herweh, CAP
Southeast	Col. Oscar K. Jolley, CAP
North Central	Col. William H. Ramsey, CAP
Southwest	Col. Luther C. Bogard, CAP
Rocky Mountain	Col. Frank L. Swaim, CAP
Pacific	Col. Howard Brookfield

WING COMMANDERS (as of June 1, 1974)

WING	COMMANDER
Alabama	*LTC Harry J. Howes
Alaska	Col. James V. Brown, Jr.
Arkansas	Col. Bob E. James
Arizona	Col. Eugene G. Isaak
California	*LTC Warren J. Barry
Colorado	Col. Thomas G. Patton
Connecticut	Col. Joseph B. Witkin
Delaware	Col. Louisa S. Morse
Florida	*LTC Henri P. Casenove
Georgia	Col. Richard A. Naldrett
Hawaii	*LTC Thomas S. Evans
Idaho	*LTC Mary C. Harris
Illinois	*LTC Robert H. Wilson
Indiana	Col. James N. Mahle
Iowa	Col. William B. Cass
Kansas	Col. Arlyn F. Rowland
Kentucky	Col. John F. Price
Louisiana	Col. William H. Cahill
Maine	Col. Richard T. Davis
Maryland	Col. Stanley F. Moyer, Jr.
Massachusetts	Col. Carl J. Platter
Michigan	Col. Edward L. Palka
Minnesota	Col. John T. Johnson
Mississippi	Col. John A. Vozzo
Missouri	Col. Donald N. Fulton
Montana	*LTC David D. Smith
National Capital	*LTC Charles X. Suraci, Jr.
Nebraska	Col. David P. Mohr
Nevada	Col. Joseph Ferrara
New Hampshire	*LTC Philip M. Polhemus
New Jersey	Col. Frederick S. Bell
New Mexico	Col. Richard A. Damerow
New York	Col. Paul C. Halstead
North Carolina	Col. Ivey M. Cook, Jr.
North Dakota	Col. Erling A. Nasset
Ohio	Col. Gerald M. Tartaglione
Oklahoma	Col. Johnnie Boyd
Oregon	Col. Roy G. Loughary
Pennsylvania	Col. A. A. Milano
Puerto Rico	Col. Rodolfo D. Criscuolo
Rhode Island	Col. Edgar M. Bailey
South Carolina	Col. E. Lee Morgan
South Dakota	Col. Eugene U. Pluth
Tennessee	Col. William C. Tallent
Texas	Col. Joseph L. Cromer
Utah	Col. Larry D. Miller
Vermont	Col. Joseph L. Roemisch
Virginia	*LTC Randolph C. Ritter
Washington	Col. Kenneth K. Kershner
West Virginia	Col. Robert E. Gobel
Wisconsin	Col. Ben D. Silko
Wyoming	*LTC Albert D. Lamb

*Indicates Interim Commander

GLOSSARY of aviation terms, acronyms and abbreviations in this volume

AAF	Army Air Field
AAF	Army Air Forces
AAC	Army Air Corps
ADF	automatic direction finder
AEC	Atomic Energy Commission
AFASC	Air Force Academy Survival Course
AFB	Air Force Base
AGL	above ground level
ALNOT	alert notice
AM	amplitude modulation
ARS	Air Rescue Service
ARRS	Aerospace Rescue and Recovery Service
ATCFC	Air Training Command Familiarization Course
CAA	Civil Aeronautics Authority
CADS	Civil Air Defense Services
CAP	Civil Air Patrol
CAP/USAF	Civil Air Patrol/United States Air Force
CCC	Civilian Conservation Corps
CD	civil defense (generic) Civil Defense (proper)
CONAC	Continental Air Command
COS	Cadet Officer School
CRAF	Civil Reserve Air Fleet
CSC	Christian Encounter/Spiritual Life Conference
DCPA	Defense Civil Preparedness Agency
DOD	Department of Defense
ELT	emergency locater transmitter
EOC	emergency operations center
ES	emergency services
FAA	Federal Aviation Administration
FCC	Federal Communications Commission
FCDA	Federal Civil Defense Administration
FM	frequency modulation
FSS	Flight Service Station
GHQAF	General Headquarters Air Force
HF	high frequency
IACE	International Air Cadet Exchange
IFR	instrument flight rules
IGY	International Geophysical Year
ITU	International Telecommunications Union
MARS	Military Affiliate Radio System
MAYDAY	international voice distress call
MC	mission coordinator
MCO	mission control officer
MSOP	Medical Services Orientation Program
NAEA	National Aerospace Education Association
NAEC	National Aerospace Education Council
NASA	National Aeronautics and Space Administration
NDC	National Drill Competition
NEC	National Executive Committee
NB	National Board
OCD	Office of Civilian Defense
OES	Office of Emergency Services
POL	petroleum, oil, lubricants
RAC	Rural Area Command
RACES	Radio Amateur Civil Emergency Services
RCC	rescue coordination center
REDCAP	actual CAP search and rescue mission
ROP	Radio Operator Permit
RT	radio telephone
SALT	Strategic Arms Limitation Talks
SAR	search and rescue
SARCAP	CAP search and rescue training mission
SARDA	State and Regional Defense Airlift Plan
SCATANA	Security Control of Air Traffic and Navigation Aids
SFOC	Space Flight Orientation Course
SMC	search mission coordinator
UHF	ultra high frequency
USAF	United States Air Force
USCGS	U.S. Coast and Geodetic Survey
VFR	visual flight rules
VHF	very high frequency
VHF-DF	very high frequency direction finder
VOR	very high frequency omnidirectional radio range
WASP	War Service Programs

Index

T

U

V

W

Y

Z